# DEMULSIFICATION

INDUSTRIAL APPLICATIONS

## SURFACTANT SCIENCE SERIES

CONSULTING EDITORS

*MARTIN J. SCHICK*
*Diamond Shamrock Corporation*
*Process Chemicals Division*
*Morristown, New Jersey*

*FREDERICK M. FOWKES*
*Department of Chemistry*
*Lehigh University*
*Bethlehem, Pennsylvania*

Volume 1: NONIONIC SURFACTANTS, edited by Martin J. Schick

Volume 2: SOLVENT PROPERTIES OF SURFACTANT SOLUTIONS, edited by Kozo Shinoda *(out of print)*

Volume 3: SURFACTANT BIODEGRADATION, by Robert D. Swisher *(out of print)*

Volume 4: CATIONIC SURFACTANTS, edited by Eric Jungermann

Volume 5: DETERGENCY: THEORY AND TEST METHODS *(in three parts)*, edited by W. G. Cutler and R. C. Davis

Volume 6: EMULSIONS AND EMULSION TECHNOLOGY *(in two parts)*, edited by Kenneth J. Lissant

Volume 7: ANIONIC SURFACTANTS *(in two parts)*, edited by Warner M. Linfield

Volume 8: ANIONIC SURFACTANTS—CHEMICAL ANALYSIS, edited by John Cross

Volume 9: STABILIZATION OF COLLOIDAL DISPERSIONS BY POLYMER ADSORPTION, by Tatsuo Sato and Richard Ruch

Volume 10: ANIONIC SURFACTANTS—BIOCHEMISTRY, TOXICOLOGY, DERMATOLOGY, edited by Christian Gloxhuber

Volume 11: ANIONIC SURFACTANTS—PHYSICAL CHEMISTRY OF SURFACTANT ACTION, edited by E. H. Lucassen-Reynders

Volume 12: AMPHOTERIC SURFACTANTS, edited by B. R. Bluestein and Clifford L. Hilton

Volume 13: DEMULSIFICATION: INDUSTRIAL APPLICATIONS, by Kenneth J. Lissant

OTHER VOLUMES IN PREPARATION

Volume 14: SURFACTANTS IN TEXTILE PROCESSING, by Arved Datyner

# DEMULSIFICATION

## INDUSTRIAL APPLICATIONS

*KENNETH J. LISSANT*
*Director of Advanced Research*
*Petrolite Corporation*
*St. Louis, Missouri*

MARCEL DEKKER, INC. New York and Basel

Library of Congress Cataloging in Publication Data

Lissant, Kenneth J
Demulsification : industrial applications.

Surfactant science series ; v. 13)
Includes bibliographical references and index.
1. Demulsification I. Title. II. Series.
TP156.D42L57 1983 660.2'84292 82-23488
ISBN 0-8247-1802-X



MARCEL DEKKER, INC.

270 Madison Avenue, New York, New York 10016

Current printing (last digit):
10 9 8 7 6 5 4 3 2 1

PRINTED IN THE UNITED STATES OF AMERICA

# Preface

This book was written for those technical persons who have been asked to solve a demulsification problem and have found little help in the existing literature. It does not contain extensive mathematical treatments although references are cited to mathematical treatments where possible. It tries to show the breadth of the areas where demulsification is of concern and to outline general approaches to the problems. It points out the sort of information that must be obtained before an effective demulsification program can be selected or designed. I wish that a nice, neat formula could be given for the solution of every demulsification problem. In this book an attempt is made to show why such a formula does not exist and how existing knowledge may be utilized to solve demulsification problems effectively.

In the first chapter the scope of the field is delineated and the prevalence of unrecognized emulsion problems is described. A brief outline of emulsion terminology and theory is given with references to more detailed sources.

In the second chapter a method is outlined for the identification and testing of emulsions. Standard tests and references are given. The next two chapters address specific emulsion types, oil-in-water and water-in-oil. Since demulsification is of such importance to the petroleum industry, chapter 5 is devoted to this area. A final chapter details several new technologies and an example of the approach to a specific problem, wool scouring.

This format has been chosen to aid the investigator in resolving a specific demulsification problem. First, the system must be defined by appropriate tests selected from those described in Chapter 2. Then, depending on the type of emulsion involved (oil-in-water or water-in-oil), reference can be made to the specific approaches outlined in Chapters 3-5. Finally, some of the newer approaches described in Chapter 6 may be able to solve problems that cannot be solved by conventional methods.

It is hoped that the present book will be a useful addition to our previous two books on emulsions and emulsion technology in this series and that the reader can use all three to become acquainted with all aspects of both making and breaking emulsions throughout many industries.

I wish to thank the management of the Petrolite Corporation for encouraging me to write this book and to express my deep appreciation to my wife Ellen, who did all the art work.

*Kenneth J. Lissant*

# Contents

PREFACE i

1 THEORY 1

Introduction 1
Nature of Interfaces 2
Types of Emulsions 4
Sedimentation 5
Brownian Movement 13
Coalescence 16
Emulsifying Agents 21
Phase Diagrams 25
Geometry 30
Settling 33
Summary 35

2 TESTING 39

Introduction 39
Emulsion Types 40
Components and Amounts of Each Phase 41
Variability 43
Particle Size Distributions 44
Dispersion Mechanisms 47
Particle Charge Testing 52
Flocculation and Creaming, General 53
Jar Tests 53

Bench Scale Tests 55
Summary 60
References 60

3 DEMULSIFICATION OF OIL-IN-WATER EMULSIONS 63

Introduction 63
Occurrence 64
Typical Treatment Plant 67
Cyçlone Separators 71
Centrifuges 73
"Stilling" Ponds 73
Types of Separators 74
Design Considerations 76
Flocculation 78
Specific Examples 79
Mechanical Coalescence 81
Overall Planning 81
Inversion and Phase Reversal 83
Chemicals Used in Demulsification 87
Summary 88
References 89

4 DEMULSIFICATION OF WATER-IN-OIL EMULSIONS 93

Introduction 93
Breaking Methods 94
Chemical Demulsification 99
Summary 103
References 103

5 DEMULSIFICATION OF PETROLEUM EMULSIONS 105

Introduction 105
Origin and Occurrence 105
Bottom Settlings and Water Testing 108
Gas Separators and Water Knockouts 109
Gun Barrels 110
Heater Treaters 111
Centrifuges 113
Filters 113
Other Mechanical Methods 114

Electrical Demulsification 114
Desalting 119
Chemical Demulsification 119
Bottle Tests 121
Treatment Methods 122
Reverse Emulsions 125
Summary 127
References 132

6 ADDITIONAL METHODS AND AREAS IN DEMULSIFICATION 135

Introduction 135
Electrodialysis 136
Reverse Osmosis 138
Ultrafiltration 139
Filtration Coalescence 140
Sludge Treatment 143
Scouring Effluent 144
WRONZ System 148
Lo-flo System 150
Summary 150
References 151

AUTHOR INDEX 153

SUBJECT INDEX 159

# DEMULSIFICATION

INDUSTRIAL APPLICATIONS

# 1

# Theory

## INTRODUCTION

Most of the effort spent studying emulsion systems has been directed at producing stable emulsions. The deliberate breaking of emulsions, except in the petroleum industry, has had much less study. There are many instances where we want a stable emulsion which can be broken on demand. Food emulsions should be stable on the shelf in the container but should break in the mouth to release the flavor. Certain medicinal formulations should not break in the mouth but should release the active ingredient at a designated point in the digestive tract. It may be said that the perfect emulsion is one which will remain completely stable forever but will immediately separate into its component phases at the touch of a magic wand. Some people seem willing to believe that such a magic wand exists. The breaking of emulsions and the separation of liquid phases is said to be more of an art than a science. In the first part of this book we shall try to present as much of the science as possible. We will then be in a position to explain some of the art.

Emulsion systems can be either simple or complex. Laboratory emulsions can be prepared which are well behaved and can be studies in a systematic manner. The emulsions that occur naturally or in industrial processes are often much more complex. We shall consider both the aspects which are common to all emulsions and the critical areas in which they may differ.

## NATURE OF INTERFACES

By definition, an emulsion consists of two immiscible liquid phases, one dispersed in the other in the form of drops. At the risk of being simplistic let us look at this definition closely. Why are liquids immiscible or soluble in each other, and why do small pieces of one liquid dispersed in another take the form of drops?

The answers to both of these questions lie in the same fundamental conditions. Figure 1.1 depicts two liquids in contact at an interface. The molecules of one phase are represented by circles and of the other phase, by polygons. Now consider the forces acting on one molecule, polygon A, at the interface.

A is attracted equally by all molecules around it in the plane of the interface. It is pulled into its own bulk phase by all of the like molecules further in the phase. It is pulled into the other phase by the attractive forces of the unlike molecules. If the attraction of the unlike molecules is greater than that of the like molecules, molecule A moves into the other phase. It is soluble in that phase. As more and more, A-Type, molecules move into the

Figure 1.1 Stylized liquid–liquid interface.

other phase, the balance of the forces may change and a point may be reached where there is no more net change of molecules across the interface. The other phase is now saturated with A. The same sort of process is also happening to B-type molecules with respect to A phase. Thus, two liquids in contact will eventually become mutually saturated with each other. This process results in the composition of each of the phases approaching a common value. If this process continues until the two phases are identical, the interface disappears, and the liquids are miscible.

Once mutual saturation has been reached, molecules of A at the interface are subject to a net force pulling them into the A phase and, likewise, molecules of B tend to move into B phase. A tendency to remove molecules from the interface results. Thus, the interface assumes a form where its area is minimal. Dispersed drops tend to become spheroidal to minimize the interface. This tendency toward minimum area is referred to as interfacial tension.

It should be noted here that the interface behaves in some ways as though it were a stretched membrane. It assumes much the same geometrical shapes that a membrane would, and it is too easy to begin to think of it as a membrane. It should never be forgotten that in an interface between two pure liquids the forces which minimize the area are all directed at right angles to the interface. They are not directed along the surface as would be the case in a stretched membrane.

Since one sphere has less surface area than two smaller spheres each of half the volume, drops of liquid tend to coalesce when they come into contact. This means that if one pure liquid is dispersed in another as small drops, an unstable system results. The drops will rise or settle, depending on density differences, and as they come into contact they will coalesce. Eventually, the system will separate into two layers. Suppose a third substance is dissolved in one of the phases. As molecules of this substance come to the interface, they can pass into the other phase, return to the bulk phase, or remain at the interface. If two liquids are immiscible, they are usually different enough that materials that are soluble in one will not be as soluble in the other. Many materials will be only sparingly soluble in either phase. Thus, materials with a limited solubility will tend to concentrate at the interface. Suppose that we have a system consisting of n-decane and water. We introduce a small amount of decanoic acid. This material will be soluble in the oil phase, but the carboxyl group will be attracted to the water phase. If we add a small amount of sodium hydroxide to the water phase, the sodium salt of the acid will form at the interface, and the sodium salt will be more soluble in the water than was the free acid. There will be a tendency for the molecules of sodium decanoate to concentrate at the interface and to align themselves so that the hydrocarbon portion of the molecule is in the oil phase and the ionic head in the water phase. The presence of these molecules at the interface will displace solvent molecules into the bulk of the phase and reduce the interfacial tension. This will lessen the tendency of the drops to coalesce. The presence of a charged particle at the interface will result in a repulsive force between drops and will also help prevent coalescence. The presence of the material at the interface will, therefore, tend to stabilize the dispersion.

## TYPES OF EMULSIONS

Even if we have the necessary components, we may not have an emulsion. One of the liquids must be broken into drops and dispersed into the other. This is a physical process and must be done properly if a stable emulsion is to form. It is very important to remember that emulsions are "constructs." They are a special physical arrangement of components. It is this configuration which constitutes an emulsion. It is general practice to divide all liquid phases into two broad classes, "water" and "oil." All aqueous or water-miscible phases are classified as water. Most others are classified as oil. This is admittedly a rather indefinite approach, but in most cases it works quite well. Because either of the two immiscible phases, an oil phase and an aqueous phase, can be the external phase two types of emulsion are possible. These are referred to as oil-in-water (o/w) and water-in-oil (W/O). Bancroft [1] states, "So far as I can see it is theoretically possible to take any pair of nonmiscible or partially miscible liquids and to make two series of emulsions over the whole range of concentrations by adding suitable third substances."

Actually, oil-in-water emulsions are so common that they are often implied when the term emulsion is used. In 1888 Quinke [2] said, "An emulsion consists of a large number of spherical drops of fat dispersed in a water liquid." Emulsions of the water-in-oil type are referred to as "invert" emulsions. On the other hand, in the petroleum industry where water-in-oil emulsions are common, oil-in-water emulsions are called "reverse" emulsions.

We are concerned in this book with the breaking of emulsions and the recovery of the phases. Since emulsions are "constructs." all that has to be done is to disrupt the structure of the system, It will then fall apart into its component phases. How can this be done? The specific answer depends on our purpose in breaking the emulsion. Which phase is to retained, which discarded? If we are removing small traces of oil from an effluent stream, we may not need to get the oil in a fully recoverable form. It may only be necessary to segregate it with a minimum of water and to be sure that the effluent water is oil-free. If the effluent water contains a small amount of an oily material that is toxic or expensive, then the recovery of the oily portion may be our prime concern.

Many emulsions that must be dealt with in industry are much more complex than the simple, two-liquid systems we have been discussing. Either or both of the liquids may contain additional dissolved substances; if a liquid is saturated with respect to a solid, an excess of solid may be present. Insoluble solids of many kinds can be involved as well as dissolved and entrained gases. To further complicate matters, the drops of dispersed phase may themselves contain droplets of the external phase or another liquid dispersed within them. Thus, we get water-in-oil-in-water and oil-in-water-in-oil multiple emulsions. It is clear that each demulsification problem has to considered separately. However, a number of general considerations are common to all problems.

## SEDIMENTATION

The breaking of an emulsion involves two processes, sedimentation and coalescence. The drops of dispersed phase settle (or rise), come into contact with each other, and coalesce. Clearly, the drops cannot coalesce unless they come together. Even then they may not coalesce unless certain conditions are met. Thus the two processes, flocculation and coalescence, are essential and interdependent.

We will first look at settling and flocculation. Consider some simple theoretical situations. First, how far apart are the drops in an emulsion? This will depend on the size of the drops and the amount of internal phase. A cube of liquid 10 cm on a side will contain 1 liter of liquid. Replace 1 $cm^3$ of the liquid with 1 $cm^3$ of another immiscible liquid. Initially, assume that the two liquids have the same density and thus eliminate buoyancy effects. If all the internal phase liquid is in one drop, it will have a radius of 0.62035 cm. In a large system of 10-cm cubes it would be 8.759 cm from its nearest neighbor. This is at a concentration of 0.1% internal phase.

If we now divide the cube equally into eight smaller cubes and break the large drop equally into eight smaller drops, each at the center of a smaller cube, the drops will have a radius of 0.3102 cm and be 4.3796 cm from their nearest neighbor. Continuing this process of subdivision we have

$$\text{Radius of a single drop} = R = \left(\frac{3V}{4\pi\,(8)^n}\right)^{1/3} \qquad (1)$$

$$\text{Distance between drops} = D = \frac{10}{2^n} - 2R \qquad (2)$$

$$\text{Number of drops} = N = 8^n \qquad (3)$$

where n is the number of subdivision steps and V is the volume of internal phase, expressed in cubic centimeters, in one liter of external phase. All dimensions are in centimeters.

Tables 1.1 to 1.4 show what happens if we start with 1, 10, 100, and 500 ml of internal phase. In Figure 1.2 all these data and some intermediate values are plotted on logarithmic scales. A line at $10^{-4}$ cm or 1 μm is drawn to emphasize the colloidal region. From the tables and the graph we can see that as the degree of subdivision increases, that is as the particle size gets smaller, the

**Table 1.1** Average Drop Spacing at 0.1% Internal Phase

| Step | Number of drops | Radius (cm) | Separation (cm) |
|---|---|---|---|
| 1 | 8 | 0.3102 | 4.3800 |
| 2 | 64 | 0.1551 | 2.1900 |
| 3 | 512 | 0.0775 | 1.0950 |
| 4 | 4,096 | 0.0388 | 0.5470 |
| 5 | 32,768 | 0.0194 | 0.2737 |
| 6 | 262,144 | 0.0097 | 0.1369 |
| 7 | $2.1 \times 10^6$ | 0.0048 | 0.0684 |
| 8 | $1.7 \times 10^7$ | 0.0024 | 0.0342 |
| 9 | $1.34 \times 10^8$ | 0.0012 | 0.0171 |
| 10 | $1.07 \times 10^9$ | 0.0006 | 0.0086 |

Table 1.2 Average Drop Spacing at 1% Internal Phase

| Step | Radius (cm) | Separation (cm) |
|---|---|---|
| 1 | 0.6682 | 3.6635 |
| 2 | 0.3341 | 1.8317 |
| 3 | 0.1671 | 0.9159 |
| 4 | 0.0835 | 0.4579 |
| 5 | 0.0418 | 0.2290 |
| 6 | 0.0209 | 0.1145 |
| 7 | 0.0105 | 0.0572 |
| 8 | 0.0052 | 0.0286 |
| 9 | 0.0026 | 0.0143 |
| 10 | 0.0013 | 0.0072 |

number of particles increases as the power of eight, and the distance between individual droplets becomes smaller. Even in an emulsion containing only 0.1% internal phase, as the particle size reaches 1 μm, the spacing is about 10 μm. If we express

Table 1.3 Average Drop Spacing at 10% Internal Phase

| Step | Radius (cm) | Separation (cm) |
|---|---|---|
| 1 | 1.4397 | 2.1206 |
| 2 | 0.7199 | 1.0603 |
| 3 | 0.3599 | 0.5302 |
| 4 | 0.1800 | 0.2651 |
| 5 | 0.0900 | 0.1325 |
| 6 | 0.0450 | 0.0663 |
| 7 | 0.0225 | 0.0331 |
| 8 | 0.0113 | 0.0166 |
| 9 | 0.0056 | 0.0083 |
| 10 | 0.0028 | 0.0041 |

Table 1.4 Average Drop Spacing at 50% Internal Phase

| Step | Radius (cm) | Separation (cm) |
|---|---|---|
| 1 | 2.4619 | 0.0763 |
| 2 | 1.2309 | 0.0381 |
| 3 | 0.6155 | 0.0191 |
| 4 | 0.3077 | 0.0095 |
| 5 | 0.1539 | 0.0048 |
| 6 | 0.0769 | 0.0024 |
| 7 | 0.0385 | 0.0012 |
| 8 | 0.0192 | 0.0006 |
| 9 | 0.0096 | 0.0003 |
| 10 | 0.0048 | 0.0001 |

the separation distance D in terms of the radius of the particle (Fig. 1.3), we find that the ratio D/R is constant for each concentration of internal phase and that it approaches 0 as the phase ratio approaches $\pi/6$. This is because we have assumed a cubic configuration [3]. At a concentration of 1% internal phase the particles will be about 5.4 radii apart. At a concentration of 10% they will be only 1.5 radii apart. Thus, it is easy to see that as the particle size gets smaller, the particles get closer together. Since, as the particles get smaller, the mean free path of the particles gets longer, it is clear that as the particles get smaller than 1 μm, the mean free path of each particle will begin to be of the same magnitude as the average spacing. Thus, collisions are more probable, and coalescence is more likely.

Suppose we move to a case where the density of the two liquids is not the same. Now the drops will begin to settle (or rise) due to the buoyancy effects. The buoyant force is counteracted by the fluid friction which increases as the velocity increases, so a particle will reach a limiting velocity V. This velocity can be calculated by the equation

$$V = \frac{2}{9} \frac{r^2(d_1 - d_2)G}{\eta} \tag{4}$$

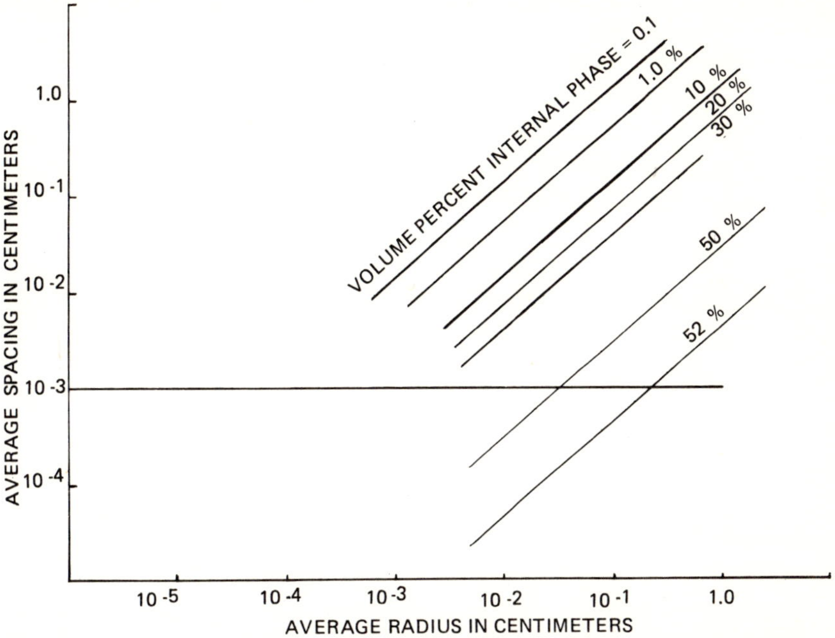


**Figure 1.2** Variation of average drop spacing with concentration.

where G is the gravitational constant, $d_1$ and $d_2$ are the densities of the two phases, (r) is the radius of the particle, and $\eta$ is the viscosity of the external phase.

Because terminal velocity is a function of the square of the radius, it drops off rapidly as the drops become smaller. Large drops, however, fall quite rapidly, so coarse emulsions will "cream" even if no coalescence occurs.

Von Smoluchowski in 1916 [4] and 1917 [5] published papers calculating the rates of collision and coalescence of particles. His equations were later refined by Weigner and Muller [6,7]. The number of drops present after a time, t, is

$$n = \frac{n_0}{(1 + \xi\beta t)^2} \tag{5}$$

where $n_0$ is the initial number of particles, $\xi$ is the collision effectiveness (fraction of collisions that result in coalescence), and

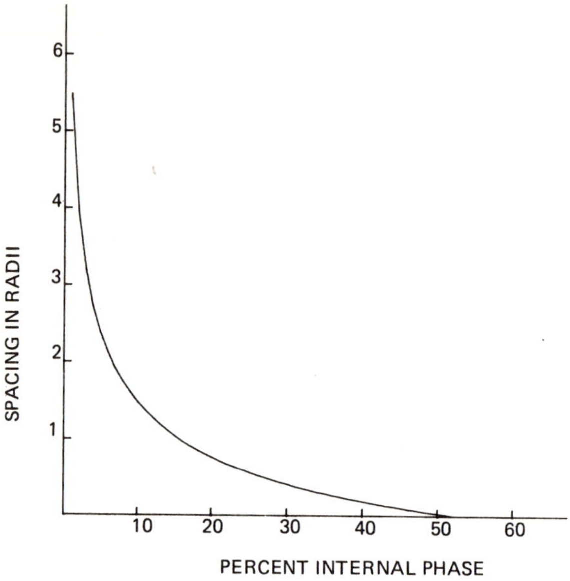


Figure 1.3 Average spacing expressed in radii.

$$\beta = DR_a n_0 \qquad (6)$$

where D is the diffusion coefficient, and $R_a$ is the *radius of action*.

It should be remembered that

$$D = \frac{RT}{N}\frac{1}{6\pi\eta r} \qquad (7)$$

where $\eta$ is the viscosity and r the radius of the drop.

Now we are in a position to study the Smoluchowski-Muller equation. If we insert the definitions of $\beta$ and D into the first equation we get*

$$n = \frac{n_0}{[1 + (\xi R_a n_0 tRT)/(6\pi\eta rN)]^2} \qquad (8)$$

---

*Be careful not to confuse the radius of action $R_a$ with the gas constant R or the radius of the drop r.

This equation has been tested experimentally by several investigators and found to be useful [7]. By setting values of the collision effectiveness and the radius of action, good plots of experimental data can be obtained. The values of $\xi$ and $R_a$ required for a fit are reasonable in most cases.

If all the constants are considered, it is apparent that the assumptions of the equation are quite acceptable. Since n is the number of particles after time t, the smaller n becomes, the more rapid the coagulation process will be. Thus, influences that decrease the value of n increase the rate of coagulation. The coagulation rate increases when:

1. The diffusion coefficient increases.
2. The particle size increases.
3. The viscosity decreases.
4. The concentration of internal phase increases.
5. The temperature increases.

Of course all these variables are not independent, so all factors must be considered in a specific case. There are a number of theoretical equations which have been developed which cover particular aspects of emulsion behavior. Stokes' law, which covers the rate of fall of drops, has been mentioned. Einstein's equation can be used to calculate the motion due to Brownian movement [8,9]. Even with the use of these equations and others it is not easy to get a clear picture of what happens as a dispersion settles and coalesces. With the aid of computers it is possible to model the behavior of an emulsion or dispersion. We can set up a model array and study the behavior of each drop. We can separate the various influences and then recombine them. When this is done a somewhat clearer picture of drop behavior begins to emerge [10].

The "universe" for our model is a cube 10 cm on a side. It contains a liquid with a density of 1 g/cm$^3$ and a viscosity of 1 cP at 273 K. Into this system are introduced various amounts of a second, immiscible liquid with a density of 1.3 g/cm$^3$.

First, consider just the effect of gravity. Stokes' law may be written in the form

$$V = \frac{2\,Gr^2(d_1 - d_2)}{9\eta} \tag{9}$$

where v is the velocity of fall, G is the gravitational constant, $d_1$ and $d_2$ are the densities of the two phases, and $\eta$ is the viscosity of the external phase.

Figure 1.4 depicts four drops placed in the model cube and allowed to fall according to Stokes' law. The drops have radii of 10, 30, 60, and 90 μm, respectively. The rate of fall has been exaggerated by a factor of 10 in Figure 1.4a to emphasize the effect of size. The positions of the drops are shown at 10 sec intervals. Since the velocity is a function of the square of the radius, the largest drop is out of the picture after less than

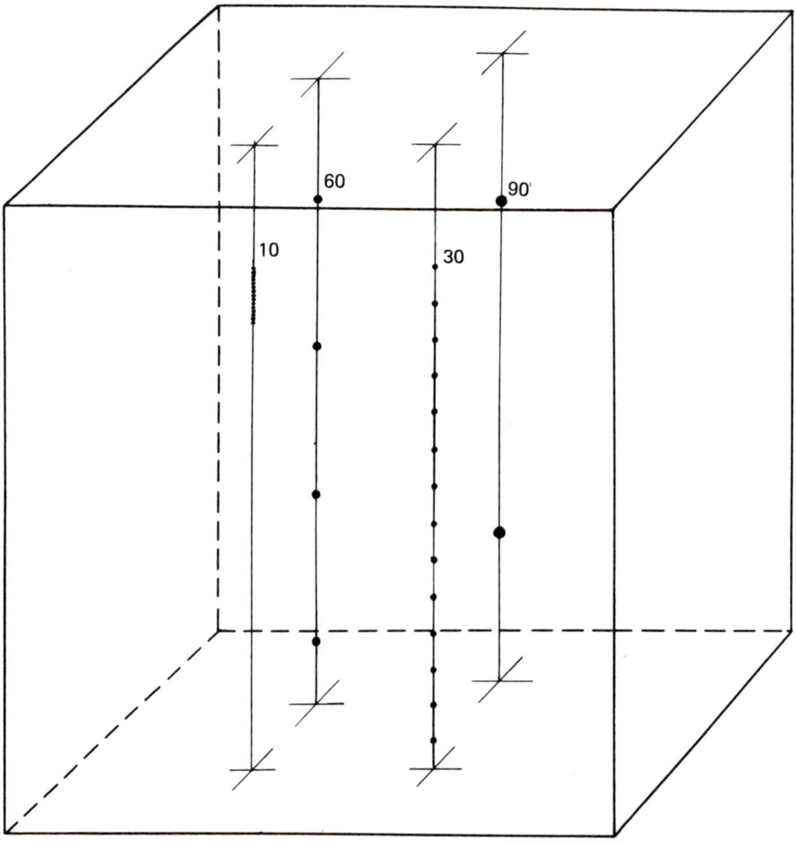


**Figure 1.4 (a)** Stokes' law falling (10×).

30 sec., while the smallest drop falls less in the total time (3 min) than the next larger does in 20 sec. Figure 1.4b shows the amount of falling on a real time scale. The smallest drop has hardly moved while the others show significant movement.

## BROWNIAN MOVEMENT

It is well known that the drops are under constant bombardment from the molecules of the external liquid and will therefore exhibit

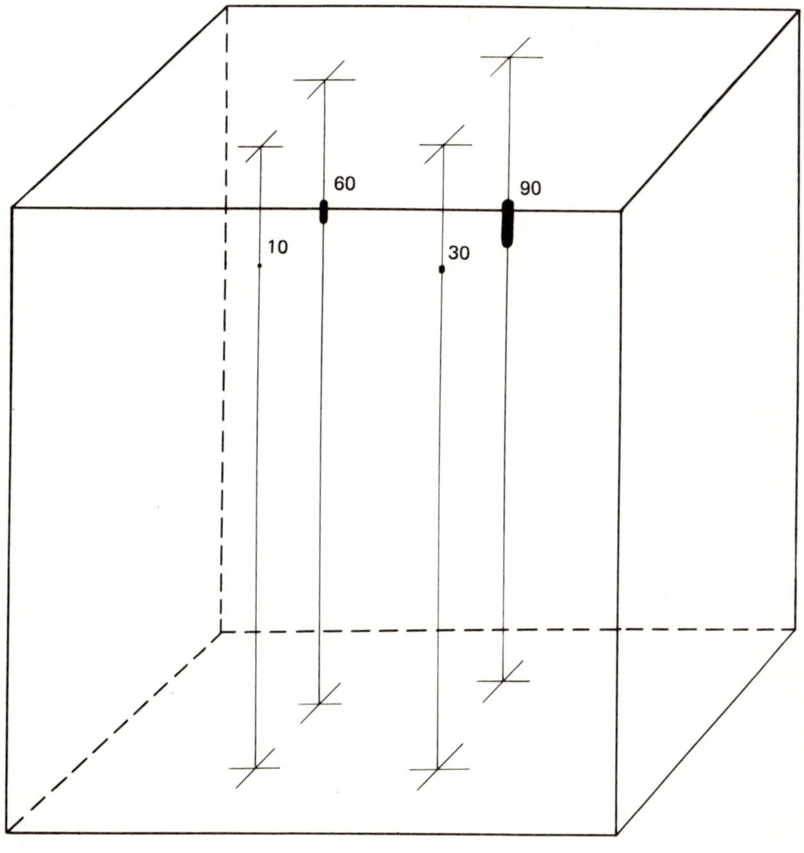


Figure 1.4 (b) Stokes' law falling.

Brownian movement. Einstein formulated an equation which allows one to calculate the average distance d a particle will travel along any axis in a given time period.

$$d = \left( \frac{RT}{N} \frac{t}{3\pi\gamma\eta} \right)^{1/2} \tag{10}$$

This relationship can now be programmed into the model. Figure 1.5a shows the same four drops with the Brownian effect

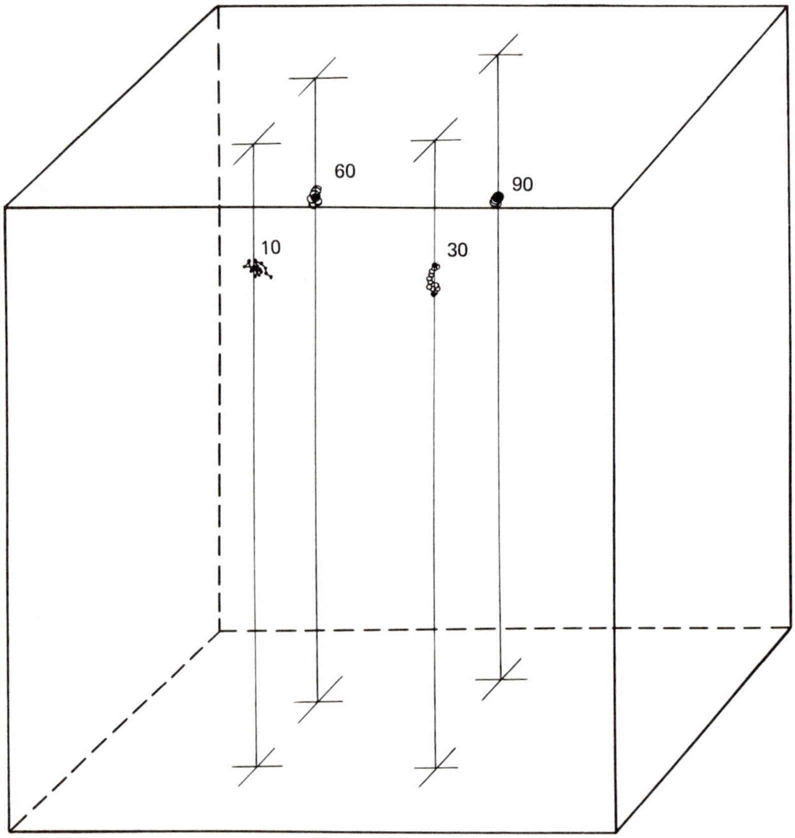


Figure 1.5 (a) Brownian movement vs. dropsize (10×).

only (not the gravity component) enlarged by a factor of ten. Here the small particles show large displacements while the large particles hardly move. Figure 1.5b shows the effect on a real scale. Even here the smallest drop shows displacement while the largest seems almost stationary. Table 1.5 displays the Stokes' effect and the Brownian effect for a range of particles between 1 and 100 μm. Note that the Brownian displacement for the 1 μm particle is over three times larger than the Stokes' effect. Under these conditions particles in the size range will not settle. Figure 1.6 shows the combined effect of both influences. The net result

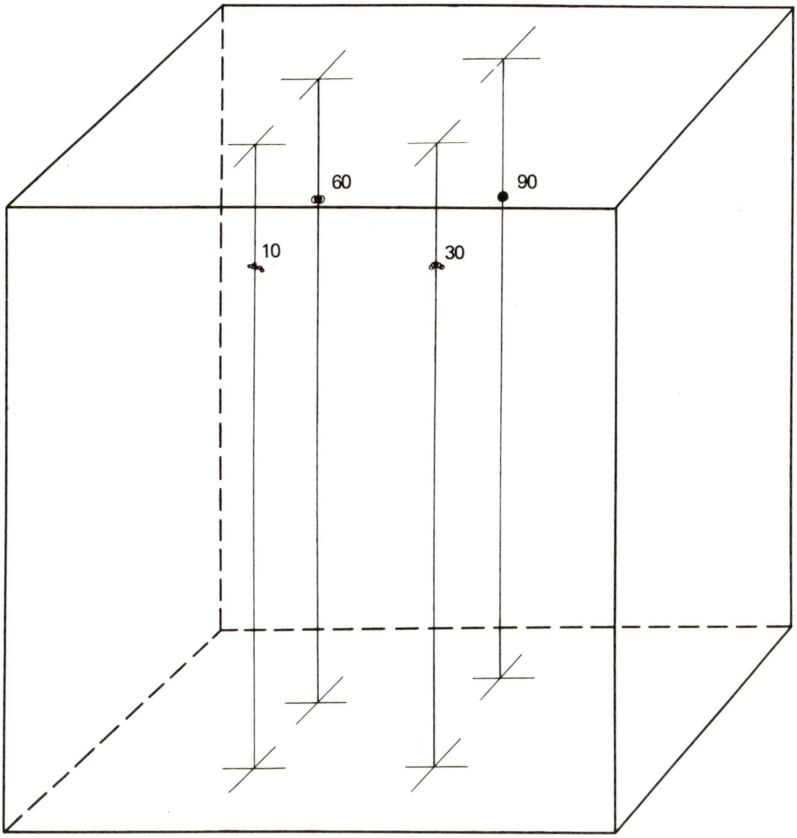


Figure 1.5 (b) Brownian movement.

Table 1.5 Particle Movement vs Drop Size

| Particle size (μm) | Stokes falling (μm) | Brownian movement (μm) |
|---|---|---|
| 1 | 0.653 | 1.9995 |
| 5 | 16.3 | 0.89 |
| 10 | 65.3 | 0.632 |
| 20 | 261.3 | 0.447 |
| 30 | 588 | 0.365 |
| 40 | 1045.3 | 0.316 |
| 50 | 1633.3 | 0.283 |
| 60 | 2352 | 0.258 |
| 70 | 3201.3 | 0.239 |
| 80 | 4181.3 | 0.224 |
| 90 | 5292 | 0.211 |
| 100 | 6533.3 | 0.1999 |

is that the larger particles fall almost without retardation while the smaller particles are kept in suspension by the Brownian movement.

This discussion assumes that neither turbulence nor convection currents are present. In actual practice any casual agitation and any thermal gradients can also help keep even some of the larger particles in suspension.

## COALESCENCE

So much for sedimentation. Now what is the effect of coalescence? Obviously, as the drops collect at the top or bottom of the system, they will be forced into contact and may coalesce. If drops collide while moving through the fluid, they may also coalesce. How may drops collide? Two drops impelled by Brownian effects may collide. This is more likely to occur between small drops since we have shown that they have larger random movements. A small drop may collide with a larger drop. Large, fast-falling drops may overtake slower drops. Drops may hit the sides of the container and

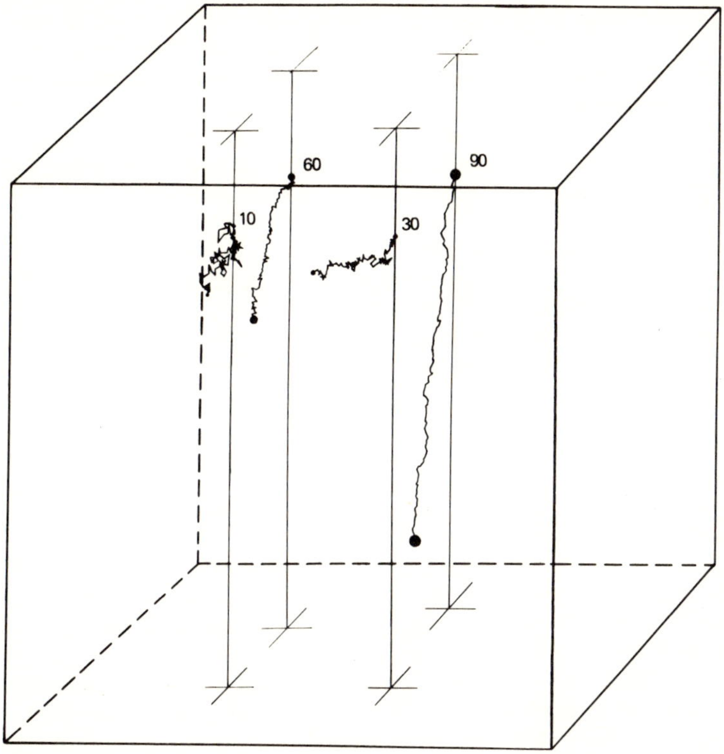


Figure 1.6 Computer simulation of effect of gravity and Brownian movement.

wet the wall, so other drops which hit the wall might coalesce with the film on the wall. All of these effects do occur, and any combination may be occurring at the same time.

Consider first the case where each time two drops touch they coalesce. This would be the case of a pure liquid dispersed in another, immiscible, unreactive liquid. Assume for now that the drops bounce off the side walls elastically. Assume also, for the simplest case, that the drops are uncharged and that no long-range attractive forces are at work. Even this very simple case can have many ramifications. Before attempting any exact mathematical description, or even a rough approximation of its behavior, it may be better to take a quick overview.

Remember that each very small drop is moving in a somewhat random walk around a point which is slowly falling. The large drops are falling rapidly in an almost straight line. One could picture the loci of the small drops as fuzzy spheres and the loci of the large drops as cylinders with a diameter equal to that of the drop. The drops of intermediate size can be viewed as decribing paths that resemble helical cylinders—the larger ones with less horizontal displacement and diameters near that of the drop, the smaller ones much more skewed and with fuzzy boundaries. Whenever the loci of two or more drops overlap in space and time, they will be likely to coalesce. The new drop will have a volume equal to the sum of the original volumes but its diameter will not equal the sum of the original diameters. All these conditions are reasonably obvious but need to be considered individually before estimating the total effect. One more factor, the concentration of the dispersed phase, has great influence on the rate of coalescence. Earlier we discussed the effect of concentration on the spacing of the drops. When the average spacing of the drops is smaller than the effective radius of the average locus, then contact becomes almost a certainty. However, as the drops coalesce, the average radius increases. The drops fall faster but subtend a smaller cross section. As the total number of drops decreases, the chances for contact also decrease. Again we have several simple factors which can add up to a complex and indeterminate situation. It will be remembered that the Smoluchowski-Muller equation for coalesence rate contained two corrections: the radius of action and the collision effectiveness. By using these two "constants" to cover the indeterminate aspects of the situation, we can fit almost any set of experimental data, but we cannot predict with precision what will happen in the next problem we encounter. This is the real difficulty in dealing with actual field demulsification cases. The best approach seems to be to look at all the factors which can apply, decide which are likely to be dominant, and then run field tests to fine tune the final solutions.

Suppose that two drops approach each other. What factors will be involved? If the drops have a charge, there will be an electrostatic effect. When the drops come very close, gravitational attraction will begin to be a consideration. If both drops are falling through a liquid, a low pressure region may develop between them, pulling them together. These and other influences may come into play in certain cases. What this seems to indicate is that for many field problems there are so many indeterminate factors that a precise mathematical statement of the interactions which will allow full predictions of events is not possible. Does this mean that we should despair of a theoretical approach? By no means. A systematic study and delineation of each of the factors is underway, and each step brings us closer to a full understanding.

Consider now how and why two drops coalesce. It is clear that one large drop has less surface area than two smaller drops, each of half its volume. If work is required to produce interfacial area, then the system will tend towards minimal area. On the other hand, as the interfacial tension approaches zero, there will be little driving force to cause coalescence. Okazawa and Bron [11] have stated that for an emulsion to be thermodynamically stable the interfacial tension must be zero and

$$\frac{\partial \Delta F^{(1)}}{\partial \eta_i^{(1)}} = \frac{\partial \Delta F^{(2)}}{\partial \eta_i^{(2)}} = \frac{\partial \Delta F^{(\sigma)}}{\partial \eta_i^{(\sigma)}} \qquad (11)$$

or

$$\Delta\mu_i^{(1)} = \Delta\mu_i^{(2)} = \Delta\mu_i^{(\alpha)} \qquad (12)$$

where $\mu_i^{(1)}$ is the chemical potential of component i in phase 1, $\mu_i^{(2)}$ is the chemical potential of component i in phase 2, and $\mu_i^{(\sigma)}$ is the chemical potential of component i in the interfacial phase.

Okazawa and Bron have published a mathematical treatment which specifies a set of conditions for thermodynamically stable emulsions [11]. Since this is one of the most recent attempts in this area, we will look at it more closely. It is known that certain emulsions can be stable for long periods of time. Friberg and others [12] have shown that certain microemulsions can be considered to be in thermodynamic equilibrium. In general, however, it has been thought that ordinary emulsions are inherently unstable.

Because emulsions contain several components and because of the presence of curved interfaces, certain constraints must be put on any mathematical treatment. Okazawa and Bron choose to omit the ionic effect on the chemical potential. This effect would have to be considered in systems containing ionic surfactants, but since nonionic surfactants can produce stable emulsions, they feel that the ionic effect can probably be treated separately [11].

In a system consisting of $N^{(D)}$ droplets of volume $v^{(D)}$ and interfacial area $a^{(D)}$, suspended in an external phase of volume $V^{(E)}$ at a temperature t and an external pressure $P_0$, Okazawa and Bron are able to show that the mechanical work done on the system can be represented as a function only of $P_0$ and the total volume V. Because of this, they choose the Helmholtz free energy rather than the Gibbs free energy for their fundamental thermodynamic functions [11]. This has also been suggested by Reiss [13].

The system is now considered to consist of two bulk phases, the continuous external phase (1) and the droplets (2), and a phase adsorbed at the interface and designated by the superscript ($\sigma$).

The mechanical work done on the system now becomes

$$dW = P_0 [dV^{(1)} + dV^{(2)} + dV^{(\sigma)}] \tag{13}$$

Note that in this approach the interfacial phase becomes three dimensional. Since

$$G = U - TS + P_0 V \tag{14}$$

we get

$$dG = - Sdt + [V^{(1)} + V^{(2)} + V^{(\sigma)}] dP_0 + \Sigma_i \mu_i^{(1)} dn_i^{(1)}$$

$$+ \Sigma_i \mu_i^{(2)} dn_i^{(2)} + \Sigma_i \mu_i^{(\sigma)} dn_i^{(\sigma)} \tag{15}$$

where $n_i^{(j)}$ is the number of molecules of component i in phase j and $\mu_i^{(j)}$ is the chemical potential of component i in phase j.

If we assume a constant temperature and a constant external pressure, a minimization of G leads to the equilibrium condition

$$\mu_i^{(1)} = \mu_i^{(2)} = \mu_1^{(\sigma)} \tag{16}$$

This condition holds regardless of the internal pressure of the drop, the total volume of the drops v, the interfacial tension, or the interfacial area. In the special case where the interfacial tension equals zero:

$$\frac{\partial \Delta F^{(1)}}{\partial n_i^{(1)}} = \frac{\partial \Delta F^{(2)}}{\partial n_i^{(2)}} = \frac{\partial \Delta F^{(\sigma)}}{\partial n_i^{(\sigma)}} \tag{17}$$

This treatment is all based on the assumptions that in the system the volume change on mixing and the compressibility are small enough to neglect.

How do we interpret these equations in a specific case? Consider a system consisting of pure water, pure n-octane, and a pure nonionic surfactant. When these three components are mixed, the water will become saturated with n-octane, the n-octane will become saturated with water, and the surfactant will distribute between the two phases. This process will attain equilibrium at a given temperature and pressure when the chemical potentials become equal. In the absence of a surfactant there will be an interfacial tension between the two phases, and work will have to be done to break one phase into droplets and disperse it into the other. Notice that either phase could become the external phase. Since work is required to produce interfacial area, the system will tend toward minimal area, and any dispersion will be unstable. If an appropriate surfactant is present, it will tend to concentrate at the interface between the two phases. This interfacial phase will consist of solvated surfactant molecules oriented towards each phase in such a way as to minimize the interfacial tension between each bulk phase and the interfacial phase [14].

## EMULSIFYING AGENTS

Surfactant molecules consist of portions that are oleophillic and portions that are hydrophylic. If these portions can be oriented so as to present the appropriate part towards each bulk phase, then the water phase will attract a portion that tends to be soluble in it and the oil phase will attract a hydrocarbon chain that is similar to it. The correctly chosen surfactant will thus effectively interpose itself between the two phases; the overall interfacial tension will approach zero, and a stable emulsion will result. In a sense we may say that we have substituted the work of orienting the interfacial phase for the work of producing interfacial area. This process is spontaneous because of the structure of the surfactant, so we may say that the energy used to produce the appropriate structure in the synthesis of the surfactant is the energy that is available to stabilize the emulsion.

All this discussion has been directed toward the production of stable emulsions. We are concerned with the destabilization of emulsions. How can we counteract the effect of the surfactant? Several approaches are possible:

1. Decompose it.
2. React with it chemically.
3. Increase the solubility of the surfactant in either bulk phase.
4. Disrupt the oriented structure of the interfacial phase.

Each approach will be considered separately.

1. *Decompose it.* Decomposition happens *in vivo* to any biodegradable surfactant. It is oxidized by the air or any other oxidizing agent; it is metabolized by a microorganism; its decomposition may be catalyzed by solid surfaces. Food emulsions may be broken by blowing them with air. In general, however, any process reactive enough to destroy the surfactant may also attack other components and is, therefore, undesirable.

2. *React with it chemically.* Chemical reaction, rather than destroying the surfactant, modifies it so that it no longer is suited for the particular system. One of the simplest of these cases is the neutralization of an ionic surfactant. If a calcium salt is added to an emulsion stabilized by a sodium soap, the calcium salt of the soap will be formed. The calcium soap is less soluble in water. The interfacial film is changed, and the emulsion may break. In both of these first two approaches the reactions are stoichiometric. The surfactant has to be dealt with mole for mole. This approach is sometimes expensive.

3. *Increase the solubility of the surfactant in either bulk phase.* Increased solubility of the surfactant is a common way of breaking emulsions. The addition of an alcohol to the system may increase the solubility of the surfactant in one of the phases and so pull it from the interface. If the aqueous phase is a brine, adding more water may pull the surfactant into the water phase.

4. *Disrupt the oriented structure of the surfactant phase.* It is postulated that the commercial demulsifiers work by disruption of the structure of the surfactant. Often these materials will be effective in extremely small amounts. They cannot be operating on a mole for mole basis. Many of these materials are not very soluble in either of the phases. This helps them concentrate at the interface. Thus, a very small amount of demulsifier can result in a fairly high local concentration of active agent at the interface. It was noted before that emulsions are constructs. We have seen that the interfacial phase owes its effectiveness to its oriented

structure. The forces that hold the interfacial phase in a particular pattern are short-range forces. If some other material can insert itself between the surfactant molecules, it increases the intermolecular distances and weakens the binding forces of the structure.

It has long been known that many surfactants are ineffective until they are present in a certain concentration. If a small amount of such a surfactant is dissolved in water it is present as single surfactant molecules. As the concentration of surfactant is increased, the individual molecules associate into polymolecular aggregates which have been named micelles. The concentration at which this aggregation begins to be noticeable is called the critical micelle concentration. As the concentration of surfactant is increased above the critical micelle concentration, the average number of molecules in each micelle increases, and the configuration of the micelle may change. It is thought that the reason that the micelles form is the same reason that the material concentrates at the interface. Part of the molecule is water-soluble, and part is oil soluble. Thus it is not really compatible with either phase. But if several molecules can associate so that the oleophylic portions are on the inside, then the water will see only hydrophilic moieties, and the micelles will be more soluble in water than the individual molecules. The reverse situation can also occur. The micelle can form with the hydrophilic portions on the inside and thus be oil-soluble. Indeed, it is possible to "dissolve" moderate amounts of the appropriate phase in the center of these spherical micelles and thus "solubilize" [15-17] water-in-oil or oil-in-water. There has been a running controversy for some time as to the difference between solubilization, micellar solutions, and microemulsions [18-20]. It seems that all three phenomena can occur but that the dividing line between them is not always obvious. In any event, it is clear that the actual aggregate which accounts for the effectiveness of the surfactant might not be, and often is not, the individual molecule, but rather some larger unit.

Experimentation has demonstrated the presence of several kinds of micelles. Spherical micelles [20,21] have the form shown in Figure 1.7a and b. Lamellar micelles have the form shown in Figure 1.7c-e. The existence of cylindrical, helical, and rodlike micelles has also been reported [22]. It has been shown that transitions from one type of micelle to another or the disappearance of micelles is brought about by changes in concentration, temperature, composition of the solvent phase, or the presence of cosolvents and cosurfactants [23-26].

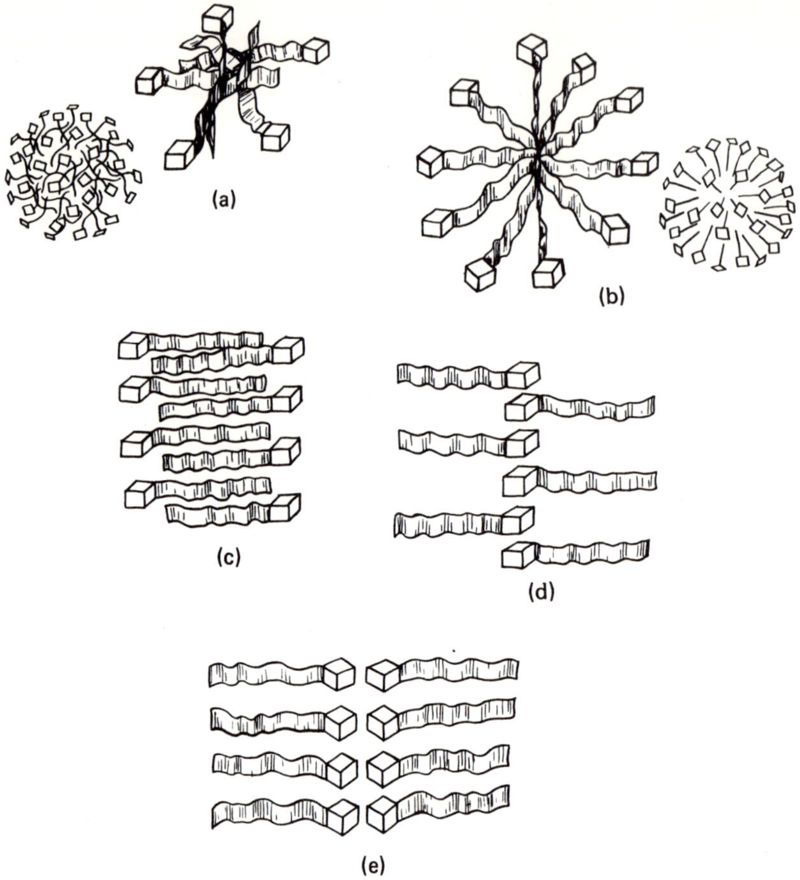


**Figure 1.7** Possible micelle configurations.

Phase diagrams of several systems have shown surprising degrees of complexity and strong dependence on several of the variables. In these phase diagrams it can be seen that the boundaries between regions are very sharp and that the introduction of very small amounts of material can throw the system from one region to another.

## PHASE DIAGRAMS

We have stated that an emulsion consists of two immiscible liquid phases. Let us review what we learned in our first physical chemistry course and recall what we know about phases, components, degrees of freedom, the phase rule, and phase diagrams [27-29].

First, let us define some terms. A phase is any physically distinct, homogeneous region of a system possessing distinct boundaries. The region does not have to be continuous. The number of components in a system is the smallest number of distinct entities needed to completely describe the system. The degrees of freedom of a system are the independent variables used to describe the system. The number of degrees of freedom is the least number of independent variables that must be specified to completely describe the system. It is usually assumed that the system is in equilibrium.

In the late 1800s, J. Willard Gibbs proposed what has been called the phase rule [30]. He said that the number of degrees of freedom of a system was equal to the number of components minus the number of phases plus two

$$F = C - P + 2 \tag{18}$$

This rule has been deduced rigorously from basic thermodynamics [31] and verified experimentally [32]. Since emulsions are polyphase systems, phase rule concepts are most useful in their study.

Methods have been worked out to record and display data concerning specific systems. One of the most useful methods for emulsion study consists of drawing phase diagrams. Begin with the system consisting of pure water. The number of components is one, so the degrees of freedom, or the variance, are always three minus the number of phases. Thus if ice, liquid water, and water vapor are in equilibrium, the system is fixed. There are no degrees of freedom. Suppose we add some n-decane. We now have two components. If the n-decane and the water were miscible there would still be one phase and the variance would be three. We would have to specify three variables such as temperature, pressure, and concentration of components to fully describe the system. Since the two liquids are not mutually soluble at standard conditions, there will be two phases and, therefore, two degrees of freedom. Thus, we had to specify "at standard conditions" in the previous sentence.

To fully describe the behavior of pure water we need specify only two variables. We can draw a graph with the temperature along one axis and the pressure along the other and fully describe the behavior of water on this phase diagram.

When another component is added, we either have to use a three-dimensional phase diagram or hold at least one variable constant. When three degrees of freedom or more exist, it is necessary to specify the concentrations of the components to fully describe the system. We therefore need a way of depicting a "composition space" which will include all possible mixtures of the components. One way of doing this is shown in Figure 1.8. If one component is present, there can be no variation in concentration, and the full system is represented by a point. All possible mixtures of two components can be depicted on a line segment. One end represents pure component A and the other end pure component B. All possible mixtures can be represented by the

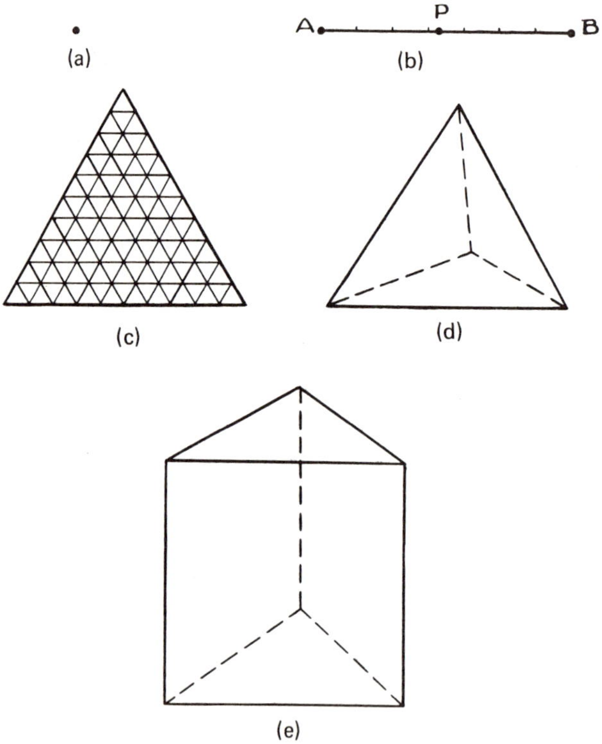


Figure 1.8 Composition spaces.

appropriate point on the line segment. A point divides the line segment into two segments, the ratio of whose lengths is proportional to the ratio of the amounts of each component. In an analogous manner all mixtures of three components can be represented within an equilateral triangle and all mixtures of four components, within a tetrahedron. It can be seen from the phase rule that even with all compositions fully specified, there may still be a need to specify two more variables.

Phase diagrams of emulsion systems may be of several types. The typical system consists of water, an oil, and a surfactant. If the oil is a pure component and the surfactant is a pure compound, we have three components. If we assume that we have two phases, three variables must be specified. If there is, in reality, an interfacial phase, yet another variable must be set. For most cases, however, we can plot test results in a triangular prism as depicted in Figure 1.8e. The triangular cross section sets the composition, and the vertical dimension represents temperature. Even if the surfactant is not one pure species, pseudoternary diagrams of this type afford a convenient form for displaying data. It should be noted that the rectangular faces of the prism represent the phase diagrams of the two component systems and the vertical slices through the prism represent plots where one component is held constant or the ratio of components is held constant. Horizontal slices are, of course, results at constant temperature.

Bearing these conventions in mind makes it easier to understand some of the papers that appear in the literature [33-36]. The use of phase diagrams makes is possible to define our area of interest somewhat more precisely. Figure 1.9 shows a typical ternary diagram. The lower left-hand corner represents 100% water phase; the top corner, 100% oil phase; the right-hand corner, 100% surfactant. Even if the water and oil phases are not pure and the surfactant is a mixture, this pseudoternary diagram is convenient. Most emulsions which require breaking contain less than 10% surfactant. In this case, we can draw a line parallel to the left-hand boundary and 10% of the distance to the surfactant apex. Only the area to the left of this line is of interest. If we are considering oil-in-water emulsions, most of them will contain less than 50% oil, so the lower left-hand corner of the diagram will be of most interest to us. This region is not usually covered in formal studies since it does not seem to contain the most interesting features. It is usually designated a two-phase region. More detailed information is available, however, in a study by Lissant and Bradley [37] using typical commercial nonionic surfactants. In particular their Figure 2c is of interest. We have taken the raw

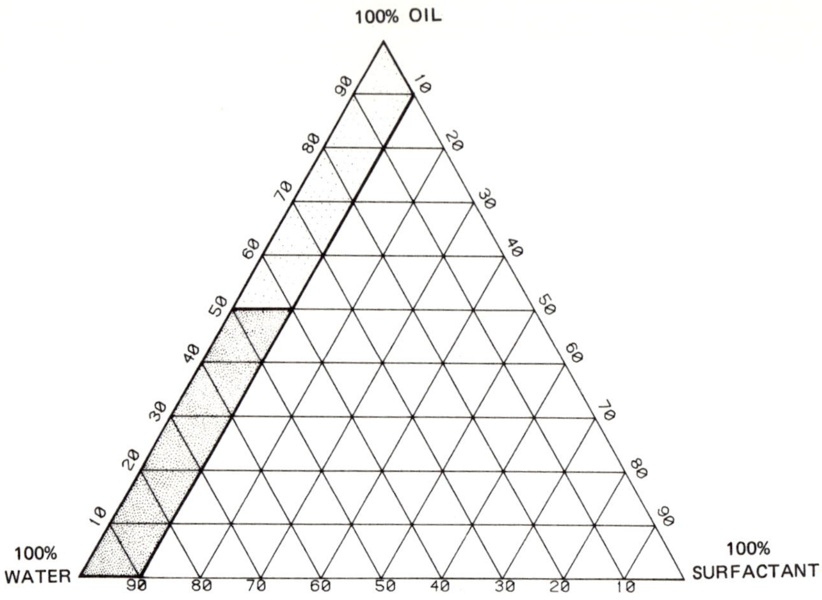


Figure 1.9 Typical ternary phase chart.

data from these tests and redrawn the region from 0% to 20% surfactant. These data are shown in Figure 1.10. Of course this system is not likely to be encountered in a real situation, but systems very similar to it may be encountered in effluent cleanup.

Suppose that we have a system consisting of water, waste oil, and surfactants that is being dumped to a sewer. Assume also that, although we do not know it, the system has a phase diagram at room temperature like Figure 1.10. The system is sampled, and an analysis reveals 67% oil, 7% surfactant, and 26% water. This composition is represented at point A. The diagram shows this to be in a two-phase region. If we assume that the surfactant is not appreciably soluble in the oil, then we can draw a line from the oil vertex through point A. We find that this line intersects a clear one-phase region. This means that if a pure oil layer separates out, the water phase will have the composition of the point where the line hits the one-phase region. This composition still contains about 43% oil. Thus if the system is allowed to settle and the clear aqueous phase discharged, an effluent with more than 40% oil would be going to the sewer. How can this problem be resolved?

Consider the case where, instead of a composition at point A, we have one corresponding to point B. Now if the system is allowed to settle, the water phase will be a thick, anisotropic gel. This liquid-crystalline phase will entrain solids and be very hard to handle.

Suppose, however, that each system is diluted with an equal amount of water. Doing so will move point A to C and point B to D. Now, upon settling, a clean oil phase and an aqueous phase containing no oil can separate out. This is one way a phase diagram can be used to understand the characteristics of a system. It will also be found that as the temperature is raised or lowered, the shape and size of the various regions in the phase diagram will change. These changes partially explain why heating a system may help in effecting a separation.

Whether the surfactant phase of an emulsion is viewed as a complex micellar region or the interface is viewed as a large unrolled micelle, we can begin to see how a very small amount of an added material can change the phase type or cause the disappearance of the interfacial film. Unfortunately, it is not possible to

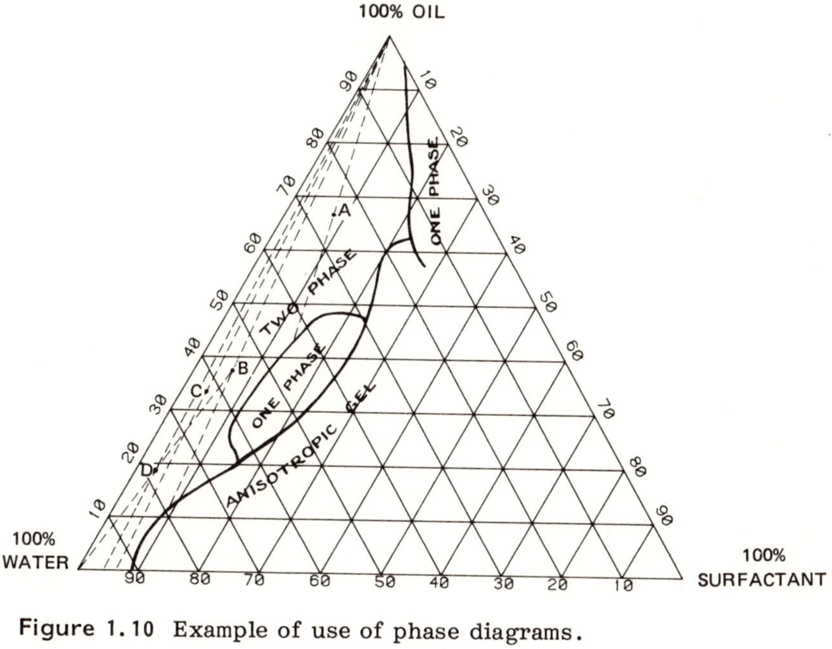


Figure 1.10 Example of use of phase diagrams.

run phase diagrams for all systems that one deals with in industry. Nor is it possible to predict what material will be most effective in each specific case. We can, however, draw on the accumulated experience of others in each area. Thus, we limit our choices and increase our effectiveness.

It should also be emphasized that in all cases the real answer to an industrial problem may be the prevention of the formation of an emulsion rather than its resolution.

One more bit of semantics should be reviewed. We should really consider an emulsion "broken" when it has been resolved into two compact phases. An emulsion is said to have "creamed" when the disperse phase has settled (or risen to the top) without the individual drops coalescing or aggregating. An emulsion has "flocculated" when some of the individual drops have associated into aggregates without coalescing. These aggregates may be clusters, chains, clumps, etc. In many cases flocculation may precede breaking. Creaming and flocculation may occur together. Thus to break an emulsion, we may first centrifuge it to hasten creaming and then cause the crowded drops to coalesce. In other cases flocculation may be promoted, thus increasing the size of the aggregates which in turn speeds up creaming. It is easy to see that all three of these processes can occur at the same time. An attempt should be made to distinguish between them and to capitalize on them. However, there is a danger in using this approach. Creamed layers or flocculated layers can sometimes be harder to break than the original emulsion. Often one is called on to solve an emulsion problem and is sent a sample of the emulsion which turns out, on further inquiry, to be a creamed or flocculated layer from the original emulsion. The client says, "We have been able to get most of the oil (or whatever) out, but we need to get out the rest." It can be difficult to get them to go back to the place where the problem started when, "We have most of it solved."

## GEOMETRY

Another factor which plays a significant part in the resolution of emulsions is the geometry of the system. Consider an emulsion to be a cloud of spheres distributed through a homogeneous medium, and now look at the basic geometry involved. This has been discussed in detail elsewhere and will not be fully developed here. The case for packing of monodisperse spheres has been considered by several investigators [38-42]. As a dilute emulsion creams, it becomes more and more concentrated; that is, the phase ratio increases. When the phase ratio reaches a little over 0.52 by volume, if all the drops were of the same size they could be arrayed each in the center of a cube. If the phase ratio exceeds this point, the drops cannot move past each other without touching. It has been found that many emulsions break at this point. It used to be said that a 50/50 phase ratio the emulsion couldn't make up its mind which way to go. Unless there is a strong interfacial film, as the drops are forced into contact, they coalesce.

As the phase ratio increases above 0.52, the drops must deform to pass. The effective viscosity of the system increases and becomes non-Newtonian. When the phase ratio reaches 0.74 . . . (the closest packing of spheres), most emulsions become unstable. For phase ratios above this point the drops must either be deformed or polydisperse or both. The mathematical treatment of such situations is somewhat more complex. Some sort of simplifying assumptions nearly always have to be made. Dodds has worked out expressions for the amount of free space based on the assumption that all spheres touch their neighbors [43]. He admits that this is not the case in actual situations. Rogers [44] has rigorously shown that the theoretical limit for the packing of equal spheres is not larger than 0.828 and, depending on how many of the spheres touch, the packing may be represented by a lower number. Actually, it is impossible for the monodisperse packing density to exceed $\pi/\sqrt{18} = 0.7404$ [45,46].

Mason has made studies of the shape of the pore space in such packings [47]. Others have also made studies with unequal spheres [48-50].

If the spheres are crowded so that they touch and flatten, they eventually approach polyhedral shapes. Thus, if emulsions possess phase ratios much above 0.80, the drops have to be deformed, and the configurations take on forms similar to that seem in foams. What are these shapes?

To begin to answer this question we must look at the forces involved. It would seem reasonable that the surface area should be kept at a minimum and that the angles between two edges at a vertex should be kept at a maximum. Observation of the degeneration of a foam will disclose that whenever more than four edges come together at a vertex another bubble slides to the vertex and eliminates it. We can now ask what shapes will fill space(tessellate) and meet these conditions. Only one fully regular shape, the cube, tessellates in three dimensions. In this configuration six edges come together at a vertex, so the cube is an unlikely candidate. Two semiregular figures are possibilities. The rhombic dodecahedron and the truncated octahedron or tetrakaidecahedron. It was assumed that plant cells and foams took on the form of rhomboidal dodecahedra or a variant called the double bee cell [50]. This was one case where perfectly obvious reasoning proved to be incorrect. Matzke in 1950 [51] and Coxeter in 1958 [52] discuss the history of this matter. It seems that Buffon in Volume 4 of his *Histoire Naturelle* stated that if peas were allowed to swell in a confined vessel, they became polyhedra. He was probably referring to an experiment reported by Stephen Hales in 1727. He states that the peas, upon being compressed, formed "pretty regular dodecahedrons." Neither man said that the dodecahedrons were rhomboidal. Apparently neither actually counted the sides or really checked the shapes. Matzke repeated the experiment with the carefully selected lead shot and found that if the shot were carefully stacked before compressing, nearly regular dodecahedrona formed. If the shot were poured into the press cylinder randomly, 14-sided figures resulted. If the shot were of differing sizes, irregular shapes resulted.

We have repeated this experiment by foaming polystyrene beads into a closed vessel in the presence of silicone fluid to prevent the fusion of the beads. We find also that the average number of faces is between 13 and 14. Few of the polyhedra are prefectly regular, but the general approach to the tetrakaidecahedral tessellations is apparent. Scanning electron microscope pictures of solidified, high-internal-phase emulsions also show this general configuration [53,54].

Princen has attempted to elucidate this emulsion problem by a theoretical treatment using a cylindrical model [55]. A further extension of this work to three-dimensional emulsions has also been reported by Princen, Aronson, and Maser [56].

Thus, it seems that as an emulsion creams, the phase ratio of the creamed layer rises and, for moderately thick, creamed layers, the draining of the cream probably stops with the phase ratio between 0.80 and 0.90. The external phase is now confined to thin films between the drops and to small, roughly tetrahedral regions, between the vertices of the drops. If the films now start to collapse, the liquid in the films will withdraw into the tetrahedra and form an inverted emulsion with very small drops of the previous external phase as the new internal phase. Such an inverted emulsion may contain from 10 to 20% internal phase. It will be more fluid than the creamed layer (unless the original internal phase is quite viscous), and since the particle size is very small, it may be harder to resolve than the original emulsion. As a further complication, many chemical demulsifiers tend to stabilize emulsions of the opposite type, so they will act as stabilizers for the inverted emulsion. This may be one of the reasons that some demulsifiers are said to "overtreat."

## SETTLING

It has been shown how Stokes' law governs the settling of single drops or very dilute dispersions. In more concentrated dispersions, however, other factors take over. Consider a uniform dispersion of small droplets of oil-in-water with an initial concentration of 0.5% by volume. These droplets will begin to rise according to Stokes' law, but as they rise, the concentration of dispersed liquid will decrease in the lower part of the column and increase at the top. As the concentration of droplets increases, they will begin to interact. At first they will form loose associations into two-drop and three-drop clusters. These drops do not need to touch. They just need to approach within a distance of about two diameters. They will then act like a single larger drop and rise faster. Clusters may grow or dissipate as rising progresses. When the concentration has increased to about 3% by volume, another factor enters. As the drops rise, they displace water which has to flow downward. This return flow slows the rise of the drops. By the time the concentration has reached about 4% to 5%, the return flow has broken the dispersion into relatively homogeneous flocs, and the return flow diffuses evenly through the cloud of drops. Now the lowest clusters begin to sweep single drops and a clear line of demarcation appears between the rising drops and the fairly clear fluid. This line of demarcation slowly rises until creaming is complete.

Most of the work done in this area has been concerned with solids which, for the most part, are heavier than water. Thus, they settle to the bottom of the flocculator. The solids may aggregate into flocs, but they do not coalesce. If we want to interpret the work in terms of emulsions, we have to remember that in some cases, such as water-in-crude-oil emusions, we will get settling while in others we will get the internal phase rising and collecting at the top. This is the case with most oil-in-water emulsions. It must also be remembered that the droplets tend to assume a streamlined shape and to distort under stress. Drops may coalesce on contact or may collect into clusters without coalescing. Solids may be occluded.

In the case of solids, four classes of settling are sometimes recognized. Class one is the Stokes' law settling discussed previously. Class two settling occurs when the concentration of droplets is high enough to cause individual particles to cluster. These clusters result in more rapid separation. Class three settling begins when the backflow slows the separation. Class four settling involves a process in which the settled layer becomes more compact.

These types of settling also occur in emulsions. If the concentration of droplets becomes high enough to cause interaction, the separation rate may increase. If coalescence begins at this stage, even faster separation occurs. If coalescence is inhibited, the concentration in certain zones of the vessel may increase to the point where the backflow of external phase again slows motion. At this point faster moving clusters catch up, and smaller drops are swept along. A line of demarcation appears. This line is sometimes erroneously referred to as an interface. At the same time if the drops that have risen to the top of the vessel begin to coalesce, a layer of oil will form. (This, of course, applies to an oil-in-water emulsion with the oil less dense than the water.) If some water gets trapped in the oil layer as coalescence proceeds, these water drops will fall toward the true interface. Figure 1-11 depicts of these stages. Figure 1-11a shows the initial uniform dispersion. Figure 1-11b drops have begun to collect at the top. Figure 1-11c shows the beginning of an accelerated region. In Figure 1-11d backflow has begun to produce a line of demarcation, and coalescence has begun to develop a liquid layer on top. The true interface is obscured by concentrated, uncoalesced drops in Figure 1-11c. The final stages of the separation are

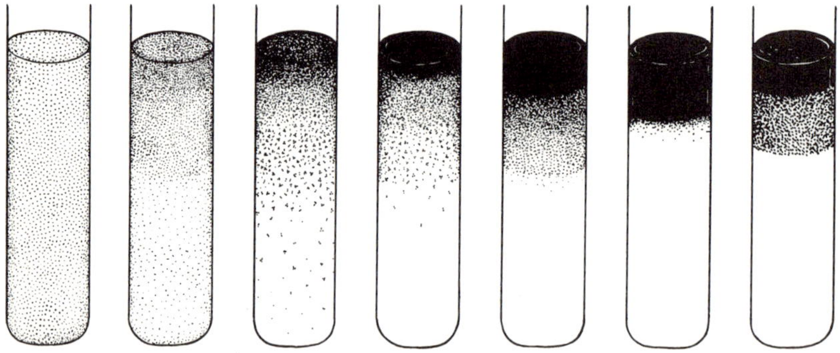

(a) (b) (c) (d) (e) (f) (g)

Figure 1-11 Hindered settling.

shown in Figure 1-11f where the last of the concentrated drops have yet to enter the interface, and a few drops of water remain to fall out of the oil phase.

In cases where coalescence is slow or incomplete or where the water drops will not coalesce out of the oil, the true interface is obscured and a "sludge layer" or "interface" develops (Fig. 1.11g). These interfaces or sludges can be very troublesome.

## SUMMARY

An attempt has been made in this chapter to cover most of the general topics that bear on demulsification. More detailed treatments of these theoretical aspects are beyond the scope of this book. However, references have been given which will provide detailed coverage. References 57-75 are included to cover various aspects that may be of interest . It is important to remember that most of these papers are aimed at explaining emulsion stability. It may be said that what we are looking for are the spectaular failures of the emulsion makers. We want to do what these people have found to result in unstable emulsions. Read in this light, some of these papers will take on a new meaning.

## REFERENCES

1. W. D. Bancroft, *J. Phys. Chem. 16*:177-233 (1912)
2. G. Quincke, *Weid Ann. 35*:589 (1888)
3. K. J. Lissant, *Emulsions and Emulsion Technology*, Marcel Dekker, New York, 1970, pp. 10-33
4. M. von Smoluchowski, *Physik Z. 17*:585 (1916)
5. M. von Smoluchowski, *Z Physik Chem. 92*:129 (1917)
6. G. Weigner and K. W. Müller, *Kolloid Z. 38*:1 (1926)
7. G. Weigner and K. W. Müller, *Kolloid Chem. Beiheft 27*:223 (1928)
8. A. Rusz, *Die Chem-Techn. Ind. 47*:317 (1951)
9. A. Einstein, *Investigations of the Theory of Browian Movement*, E. P. Dutton and Co., New York, 1926, p.32
10. I. Goodarz-Nia, *J. Colloid Interface Sci. 70*:306-319 (1979)
11. T. Okazawa and J. Bron, *J. Colloid Interface Sci. 69*:87-96 (1979)
12. E. Ruckenstein, *J. Colloid Interface Sci. 66*:369-71 (1978)
13. H. Reiss, *Methods of Thermodynamics*, Blaisdel, New York, 1965.

14. G. Brady, L. Mandelcorn, and C. A. Winkler, *Can. J. Chem. 31*:55-64 (1953)
15. J. W. Mc Bain and K. J. Lissant, *J. Phys. Colloid Chem. 55*: 655 (1951)
16. J. W. McBain and M. E. McBain, *J. Am. Chem. Soc. 58*: 2610 (1936)
17. K. Hess, *Fette und Seifen, 46*:570-73 (1939)
18. J. H. Schulaman and T. P. Hoar, *Nature, 152*:570-73 (1939)
19. J. H. Schulman and P. P. Riley, *J. Colloid Sci. 3*:383 (1948)
20. K. Shinoda, T. Nakagawa, B. Tamamuski, and T. Isemura, *Colloidal Surfactants*, Academic Press, New York, 1963, pp. 16-25
21. J. W. McBain, *Colloid Science*, D. C. Heath and Co., 1950, p. 268
22. J. W. McBain, *Colloid Science*, D. C. Heath and Co., 1950, pp. 255-261
23. P. Debye and E. W. Anacker, *J. Phys. Colloid Chem. 55*: 644 (1951)
24. E. W. Anacker, *J. Colloid Sci. 8*:402 (1953)
25. A. I. Yurzhenko and R. V. Kucher, *Kolloid Zhyr 14*:219 (1952)
26. T. Nakagawa and K. J. Kuriyama, *J. Chem. Soc. Japan, Pure Chem. Sect. 75*:73 (1954)
27. A. Findlay, *The Phase Rule and Its Applications*, Longmans, London (1906)
28. A. Findlay, A. N. Campbell, and N. O. Smith, *The Phase Rule and Its Applications*, Dover Publications, 1951
29. J. E. Ricci, *The Phase Rule and Heterogeneous Equilibrium*, D. Van Nostrand Co., New York, 1951
30. J. W. Gibbs, *Collected Works Vol. 1*, Longmans, London, 1928, pp. 55-150
31. S. Glasstone, *Textbook of Physical Chemistry*, D. Van Nostrand Co., New York, 1940, pp. 469-470
32. A. W. Adamson, *Physical Chemistry of Surfaces*, 2nd Ed., Interscience Publishers, John Wiley and Sons, New York, 1967, pp. 519-520
33. J. F. Eicke, *J. COlloid Interface Sci. 68*:440 (1971)
34. F. Harusawa, S. Nakamura, and T. Mitsui, *Colloid Polymer Sci. 252*:613-619 (1974)
35. A. V. Smirnova, A. F. Koretski, and N. A. Sokalovskaya, *Kolloid Zhur 38:* 726-730 (1964)
36. G. M. Bradley and K. J. Lissant, *J. Phys. Chem. 84*: 1567-1570 (1980)
37. K. J. Lissant and G. M. Bradley, in *Solution Chemistry of Surgactants*, (T. Mittal, ed.), Plenum, New York, 1979, pp.919ff.

38. F. Toth, *Regular Figures*, (International Series of Monographs on Pure and Applied Mathematics) Vol. 48, Macmillan, New York.
39. R. H. Brisford, *Nature (London)*, *224*:550 (1969)
40. K. J. Lissant, *J. Colloid Interface Sci.* *22*:462 (1966)
41. K. J. Lissant, *Emulsions and Emulsion Technology Part 1*, M. Dekker, New York, 1974, 10-33
42. M. Gardner, *New Mathematical Diversions*, Simon and Schuster, New York, 1966, p. 88
43. J. A. Dodds, *J. Colloid Interface Sci.* *77*:317 (1980)
44. C. A. Rogers, *Proc. London Math. Soc.* *8*:609 (1958)
45. H. S. M. Coxter, L. Few, and C. A. Rogers, *Mathematika* *6*:147-157 (1954)
46. R. P. Bambah, *Pro. Camb. Phil. Soc.* *50*:203-208 (1954)
47. G. Mason, *J. Colloid Interface Sci.* *35 (1971)*
48. J. L. Lebowitz and D. Zomick, *J. Chem. Physics* *54*:3335 (1971)
49. W. M. Visscher and M. Bolsterle, *Nature* *239*:504 (1972)
50. I. Smalley, *Mathematics Magazine*, Nov. 1963, p. 295 ff
51. E. B. Matzke, *Bulletin of the Torrey Botanical Club* *77*: 222 (1950)
52. J. S. M. Coexter, *Illinois Journal of Mathematics* *2*:746 (1958)
53. K. J. Lissant and K. G. Mayhan, *J. Colloid Interface Sci.* *42*:201 (1973)
54. K. J. Lissant, B. W. Peace, and K. G. Mayhan, *J. Colloid Interface Sci.* *47*:416 (1974)
55. H. M. Princen, *J. Colloid Interface Sci.* *71*:55 (1979)
56. H. M. Princen, M. P. Arnson, and J. C. Maser, *J. Colloid Interface Sci.* *75*:246 (1980)
57. W. J. Phillips, R. W. Graves, and R. W. Flumerfelt, *J. Colloid Interface Sci.* *76*:350 (1980)
58. R. D. Vold and K. L. Mittal, *J. Colloid Interface Sci.* *38*: 295 (1972)
59. S. Hartland and D. K. Vabra, *J. Colloid Interface Sci.* *77*: 295 (1980)
60. D. Stigter, *J. Phys. Chem.* *84*:2758-2762 (1980)
61. R. W. Flumerfelt, *J. Colloid Interface Sci.* *76*:330 (1980)
62. I. Goodarz-Nia, *J. Colloid Interface Sci.* *70*:206 (1979)
63. R. Ober and C. Taupin, *J. Phys. Chem.* *84*:2418-2422 (1980)
64. E. Ruckenstein and R. Krishnan, *J. Colloid Interface Sci.* *76*:188 (1980)
65. E. Ruckenstein and R. Krishnan, *J. Colloid Interface Sci.* *76*:201 (1980)

66. B. W. Ninhan, *J. Phys. Chem. 84*:1423 (1980)
67. A. M. Bellocq, J. Biaes, B. Clin, A. Gelat, P. Lalanne and B. Lemanceau, *J. Colloid Interface Sci. 74*:311 (1980)
68. J. T. Davis, *Proc. Int. Congress Surface Activity, London 1*:426-438 (1957)
69. A. S. Kertes and W. C. Lai, *Colloids and Surfaces 1*:197-208 (1980)
70. S. Ikeda, S. Hayaski, and T. Imae, *J. Phys. Chem. 85*: 106-112 (1981)
71. B. H. Kaye and R. P. Boardman, *Proceedings of Symposium on the Interaction Between Fluids and Particles*, Ser. #9, Inst. of Chem. Engineers, London, 1962, pp. 17-21
72. N. F. Borys, S. L. Holt, and R. E. Barden, *J. Colloid Interface Sci. 7*:526 (1979)
73. S. Ikeda, M. A. Tsunoda, H. Maeda, *J. Colloid Interface Sci. 70*:448 (1979)
74. A. Roy and G. Nemethy, *J. Phys. Chem. 76*:809 (1971)
75. Y. Moroi and R. Matuera, *J. Colloid Interface Sci. 76*:256 (1980)

# 2
# Testing

## INTRODUCTION

There are many tests that can be performed on emulsions. Most of these tests are designed to assess the quality of a particular emulsion. There are tests for stability under certain conditions, for physical properties, etc. In contrast, the tests used in demulsification are scattered throughout the literature, often under other guises.

If we want to break an emulsion, how do we start? First, we have to establish whether it is really an emulsion. How do we proceed? An emulsion must, by definition, have two liquid phases. It has an external phase and an internal, or dispersed, phase. It may have solids in either phase or at the interface. It may even have more than two liquid phases. How can such a mixture be tested?

There are several quick tests which can be run on a potential emulsion sample to see if it really is an emulsion. Visual examination is the simplest. If the sample is cloudy or opaque with visible amounts of a second liquid at the top or bottom or sticking to the sides of the container, you can be reasonably sure an emulsion is involved. If the sample is opaque with no sign of any separation of either solids or liquids, it probably is an emulsion, but it may be a slurry of finely divided solids. If it is fluid, an attempt can be made to filter it.

Now mix the sample thoroughly and pour some of it into a test tube. Immerse the test tube in a hot water bath. Remove occasionally, and shake and observe. Does separation occur? Do solids settle? Does the sample change color? Many emulsions are light brown or cream-colored. As they begin to break, they will first darken and then separate. Often the progress of the demulsification can be followed by the change in color.

If changes occur, determine if they are reversible on cooling. Sometimes phases will appear on heating or cooling but disappear on temperature reversal. Once an emulsion has broken, it may not be able to be reconstituted, even at the original temperature, without intense mixing. Sometimes attempts to reemulsify an emulsion will result in phase reversal.

Suppose you have a clear one-phase liquid which becomes cloudy on heating. If this occurs, suspect either a microemulsion or the presence of a nonionic detergent. Centrifugation, either hot or cold, may give useful information. A clear aqueous system which does not separate on heating but which shows the presence of hydrocarbons or other oils when analyzed should be checked for the presence of substantial amounts of surfactants and probably treated as a micellar solution. These systems are not extremely common. Obviously, except for microemulsions, these cases are outside the scope of this book.

If simple heating has resulted in a separation into two liquid phases, then you have an emulsion which can be resolved in this manner on any scale you wish. But heating large amounts of liquids is expensive. There may be better ways.

If a second liquid is detected, we proceed to use more sophisticated tests as decribed below.

## EMULSION TYPES

At first, an attempt will be made to determine if the external phase is oil or water. This can be done in several ways. If a drop of the emulsion and a drop of tap water are placed side by side on a glass plate or microscope slide and caused to touch by tipping or pushing with a glass rod or platinum loop, a water-external emulsion will bleed into the water drop making it milky. Similarly, an oil-external emulsion will mix with a drop of hexane, kerosene, or other oil. It is advisable to make both tests since viscous emulsions are sometimes slow to bleed into either water or oil [1-4].

Alternatively, one can test to see if the system is electrically conductive. A useful device for such tests is described in Volume 15 of the American Society for Testing and Materials, ASTM, testing manuals in test D244 [5]. It is here called a particle charge tester, and this use will be discussed later. It consists of a battery or power supply which will provide 0 to about 10 volts dc, a rheostat to control the voltage, an ammeter to measure current and an appropriate probe. For simple conductivity testing, a piece of twinlead antenna wire with the ends bared for about 3mm will serve quite well. Set the rheostat so that it gives a deflection in the middle of the ammeter scale when the probe is placed in tap water. Immerse the end of the dry probe in the emulsion and note if there is a detectable deflection. A water-external emulsion should conduct a current.

However, answers are not always clear-cut. If there is a thin layer of oil on the surface of the emulsion, it may coat the probe as it enters the sample and insulate it from the water. If the probe is water wet when it enters an oil-external emulsion, a chain of water drops may form across the gap between the wires and allow current to flow, thus giving a false reading. These tests should not be done slavishly or by rote. They require attention and interpretation [6]. Each time a new sample of emulsion is obtained, it should be tested. Some systems appear to be a water-in-oil (W/O) emulsion at one time an an oil-in-water (O/W) emulsion at another. Be sure your samples are representative and as fresh as possible. A sample which has partially separated and been remixed may have reversed its phase relationship. These occurences are common enough to be well worth checking for.

Another similar method for testing emulsion type is given by Sherman in his book *Emulsion Science* [7]. He suggests that a crystal of water-soluble dye be made to touch a drop of the emulsion on a microscope slide. If the color can be seen streaming into the drop, the emulsion is O/W. The test can be repeated with an oil-soluble dye. This method is attributed to Robertson [8]. Brilliant Blue FCF is suggested as the water-soluble dye while Sudan Red 111 and Oil Red XO have been used as an oil-soluble dye.

Some oils are fluorescent under ultraviolet (UV) light. If this is the case, a drop of emulsion can be viewed under UV light, and if distinct specks of light appear, the emulsion is O/W. If the whole field fluoresces, the emulsion is W/O.

If a drop of emulsion is placed on a filter paper and it spreads leaving a few dark spots in the middle, it is probable water external. This can be checked further by putting another drop on a filter paper that has been wet with xylene. Only an oil-external emulsion will spread.

## COMPONENTS AND AMOUNTS OF EACH PHASE

Having determined the type of emulsion, we now need to separate it and identify the components of each phase. Methods of separation that are not practical for full scale commercial use are often very useful for laboratory testing. One of the tests that has been used for years to determine the amount of oil in an O/W emulsion is the Babcock test. This test was originally developed for testing the amount of butterfat in milk. A sample of the emulsion to be tested is placed in a special "Babcock bottle" which has a long, thin, graduated neck. Sulfuric acid is added to destroy the emulsifer, and the bottle is spun in a centrifuge until the oil layer separates. The percent oil can then be read directly from the graduations on the neck of the bottle. If the amount of oil phase is above about 8%, the emulsion must be diluted before testing. Full instructions for this test can be found in several sources. ASTM test D1497 in Part 23 gives one procedure [9]. The test as it applies to milk is given in the manual of the Association of Official Analytical Chemists [10].

There are also several test procedures where the oil in a water sample is extracted with an organic solvent such as ether or chloroform. For very small amounts of oil, the material may be adsorbed by a charcoal column and extracted with an appropriate solvent. Methods of this type are described in ASTM tests D2778 and D2910 [11,12].

Similar tests are used for separating or determining the amount of water in a W/O emulsion. ASTM test D96 [13] gives the procedures for determing the amount of bottom settlings and water (BS & W) in crude oils and petroleum products. This method consists of diluting with an aromatic solvent such as xylene and centrifuging in graduated tubes. These tubes are graduated from the bottom up since the water and solids collect on the bottom.

There is also a method for determining the amount of water in oil known as the Deane-Starke method. This is described in ASTM tests D244 and D95 [14,15]. It consists of diluting the sample with benzene, toluene, or xylene and azeotroping the water into a special trap which collects the water and returns the solvent to the boiling flask. This method leaves any inorganic salts that were in the water phase in the boiling flask along with any solids.

If enough sample is available, simply heating and letting the emulsion sample stand or centrifuging it will, in many cases, effect a separation. The phases can then be analyzed by any suitable method. Infared, ultraviolet, and emission spectroscopy can be used. Standard methods of wet analysis should not be ignored. Microscopic examination will sometimes be quite revealing, particularly when solids are present.

Sometimes, if there is only a small amount of internal phase and if the droplet size is very small, the emulsion will be almost clear. It is well to remember that not all emulsions are turbid. If the indices of refraction of both phases are close enough to each other, the system will be clear or faintly smoky and still contain substantial amounts of emulsified material. Under certain conditions microemulsions can form. These systems are very stable but can be disrupted by heat, change of pH, or addition of ethanol or methanol. Microemulsions have such small droplets that the system appears perfectly transparent. It is also well to remember that not all "emulsions" are really emulsions. So-called emulsion paints, wax polish emulsions, and emulsion polymers may not be emulsions at all. They all started out as a dispersion of one liquid in another, but, either by cooling or by chemical reaction, the internal liquid was converted to a solid. It is possible to flocculate these systems, but the particles will not coalesce.

From an industrial standpoint, flocculation of suspended solids and settling of emulsions have much in common. Some industrial problems can be solved by flocculating the solids and carrying down the emulsified liquids with them. United States Patent 3,378,418 discloses a process for breaking certain emulsified fuels by contacting them with a solid that will absorb the water, thus breaking the emulsion.

So, we have confirmed the existence of an emulsion; we have determined the amounts and nature of the internal and external phases; and we have ascertained the nature of any solids that are present. We must now work out a method of breaking the emulsion and separating the phases on a practical, economical basis.

## VARIABILITY

Before one can design a system for treating a particular stream, some other facts about the system must be known. First, one should determine the degree of variability of the system. Is this a constant problem, or does it only occur at the end of a shift when equipment is washed? If cleanup is done at night, are significant volumes of wash water or other streams dumped in? Are acid or basic streams added at random intervals? Take representative samples over a considerable period and determine the degree of variation that can be expected. Does the temperature of the flow vary much from day to night or from winter to summer? Failure to take variability into account can result in a system that only works some of the time.

## PARTICLE SIZE DISTRIBUTIONS

It is known that extremely fine particles settle slowly. Therefore, it may be useful to determine the particle size distribution of the system under study. This is not done as often as it should be because, in general, particle size distributions are not easy to determine. Also, if coalescence is taking place, the distribution may change while you measure it. There are several ways that particle sizes can be determined. First, the appearance of the system will give some clues. If the drop size is above about 200μm, individual drops will be visible with the unaided eye. Particle sizes between 10 and 100μm will cause the emulsion to appear milky white; from 10 to 1μm, a bluish white; from 1 to 0.5μm, smoky gray; and below that the system will be transparent. Color is a general indication, but it does not really tell anything about the distribution of particle sizes. Detailed particle size distributions are made using methods bases on three principles. One set of methods uses microscopic examination and visually counts the drops in each size range. Another set of methods utilizes the fact that the rate of settling is a function of size. A third set of methods is based on the fact that drops of different sizes scatter light in different ways.

The oldest methods are based on visual observation. A sample of the emulsion is placed in a microscope cell. Either the drops of each size range are counted directly, as in the first attempts, or a photomicrograph is taken of a sample and a print made at a known magnification. The size of each drop shown on the print is then estimated and recorded. This is a time-consuming and tedious process. Several ingenious schemes and devices have been suggested to make things easier or quicker [16-20].

It is possible to get some idea of the particle size distribution from light-scattering data. The angle and intensity of the scattered light are functions of the size and shape of the particles as well as the refractive indices of the two phases and the wavelength of the light. The indices of refraction can be determined, and monochromatic light can be used. In monodisperse systems, we can then get the particle size, assuming that the drops are spherical. In actual emulsions there is a distribution of sizes which complicates matters. The actual measurements that are made yield plots of intensity against reflectance angle. For a single size of drop, the curve shows a series of maxima and minima. From these values the size can be calculated. Each drop size yields its own curve, and in an emulsion with a distribution of sizes, all these curves are superimposed. Details of these methods are available in several texts and articles [21-28].

Another way to determine particle size distributions is to use sedimentation data. We have mentioned Stokes' law previously. It can also be stated as:

$$D = K\left(\frac{h}{t}\right)^{1/2} \qquad \text{where} \qquad K = \left(\frac{18n}{G \cdot (d_2 d_1)}\right)^{\frac{1}{2}} \qquad (1)$$

Thus, a particle of diameter D will fall a distance h in a time t. If we have a suspension of drops, they will settle and after time t, all particles larger than diameter D will have fallen past a distance h from the surface of the suspension. If we measure the amount of suspended material at certain times and certain distances from the surface, we can calculate a size distribution. Measuring the amount of suspended material can be a problem. In one method, a metal plate is suspended in the test chamber from the arm of a sensitive balance. As the drops settle on the plate, the increase in weight is measured. In other methods, small samples are taken from different depths at measured intervals and analyzed directly for suspended material. A measurement of the amount of suspended material. A measurement of the amount of suspended material can also be obtained by determining the density of the mixture.

There is an apparatus called a "Sedigraph" [29] which is intended for particle size determinations of solid particles that can be used in some instances for measurements on emulsions. The instruction manual says, "The instrument determines, by means of a finely collimated beam of X-rays, the concentration of particles remaining at decreasing sedimentation depths as a function of time." While this instrument was not initially designed for use with emulsions, it has been tried in emulsion applications using both oil-in-water and water-in-oil emulsions. The emulsions must be stable against coalescence but must cream in order to get readings. Care must be taken that dilution of the emulsion for use in the test does not result in destabilization or further dispersion.

Another device that has been used to obtain data on the particle sizes in emulsions is the Coulter counter* [30]. This instrument operates on an entirely different principle. Again, to quote from the manufacturer's literature:

> Operation is based on electrical conductivity difference between particles and common diluent. Particles act as

*Manufactured by Coulter Electronics Inc., 590 W. 20th St., Hialeah, FL., 33010

> insulators, diluents as good conductors. The particles suspended in an electrolyte, are forced through a small aperture, through which an electrical current path has been established. As each particle displaces electrolyte in the aperture, a pulse proportional to the particle volume is produced. Thus, a three-dimensional particle volume response is the basis for all sizing, regardless of position or orientation of the particle in solution [30].

This description is obviously directed to determinations of solid particles. However, a good deal of work has been done on emulsions by several investigators. It must be remembered that, since an electrical current path is required, only emulsions of the oil-in-water type can be tested. The external phase must be conductive. Again, the emulsion must not break before a determination can be made.

In instances where the Coulter counter can be used it is a fast, convenient method. Higuchi reports tests on dibutyl phthalate-in-water emulsions [31]. Rowe describes the "Effect of Emulsifier Concentration and Type on the Particle Size Distribution of Emulsions" using a Coulter counter [32]. In 1962 Wachtel and LaMer published an article which describes the use of this instrument in detail and compares the results with those obtained by other methods [33]. Other work was reported by Schrenzel in 1966 [34]. In 1974, Lien and Phillips published a study on oil-in-water emulsions using the Coulter counter. They point out several factors which must be controlled to get dependable results [35]. They find that the apparatus and cell must be shielded from outside electric fields to minimize background noise. The manufacturer states that more recent models of the counter take this into account. The currently recommended model is the TA II with a Model PCA accessory. I have not used this model and cannot report from direct experience. Manufacturers literature recommends that for control work, a narrow band model ZB with a C1000 Chanellizer can be used. To make particle size counts either photographically or with an instrument such as the Coulter counter, it is often necessary to dilute the emulsion. Dilution can be a source of erroneous results. Some oils are appreciably soluble in water or detergent solutions, so some of the dispersed phase may be dissolved in the dilution process. Diluent solutions should be saturated with the internal phase before use if possible. For routine control work it is often desirable to retain some of the clarified external phase as a diluent. This material must not contain any demulsifers. If the dilution step results in a change in the concentration of the surfactant, false results can also be obtained.

## DISPERSION MECHANISMS

Once a particle size distribution has been obtained, how can it be used? Remember in this case we have an emulsion which we want to avoid or break. If the size distribution is changing with time, the emulsion may be breaking. If the size distribution changes when the emulsion is subjected to a certain treatment, it gives clues for designing procedures.

Consider how one liquid can be dispersed into another. What are the mechanisms that can be involved? A stream of liquid may flow into another liquid with little turbulence. The stream may be injected violently. A vapor can be mixed into a liquid and condense into droplets. The solubility of one liquid in another may change, causing precipitation to occur. Two liquids in contact may be subjected to agitation or other forms of shear. A detailed theoretical discussion of this area has been given by Clift, Grace, and Weber in their book *Bubbles, Drops, and Particles* [36].

Before going into dispersion mechanisms in any detail, it is necessary to review the various methods for expressing particle size distributions. Regardless of the method used, what is finally obtained is a count of the number of particles that fall into each of a series of size ranges. The common way to plot these data is to make a plot of radius versus number of particles. The radius is usually the abscissa. However, you will also find plots of the diameter versus the cumulative mass (or volume) percent. Both of these methods have their special uses. It is necessary to be careful that the method of presenting the data does not obscure the real meaning. Consider a system consisting of 100 drops--10 each with diameters of 10,20,30,40,50,60,70,80,90, and 100μm. A plot of this system could look like Figure 2-1a. If we plot the volume percent versus the diameter, we get Figure 2-1b. If we plot the cumulative volume percent versus the diameter, we get Figure 2-1c. These data are listed in Table 2-1. Admittedly, this is an unusual and unlikely distribution, but it may make a point. The average diameter of the particles is 55μm. There are no particles of average size. All particles within 15μm of the average constitute only 30% of the particles and only 24% of the total volume. On the other hand, removing 30% of the largest particles will remove about 74% of the total volume. But failing to remove 30% of the smallest particles will leave only a little over 1% of the material.

Now we shall look at some mechanisms whereby a liquid can be dispersed into another liquid and see what kinds of particle size distributions these mechanisms can produce. As the first case consider a large drop of one liquid suspended in another liquid. Assume the system is in zero gravity and completely at rest. The

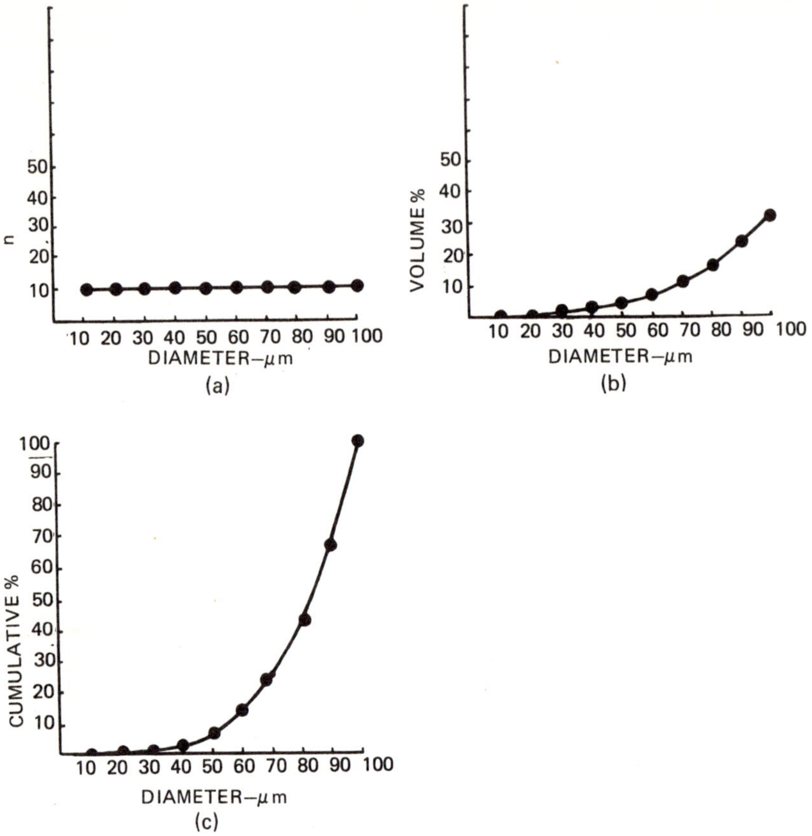


Figure 2.1 (a) Plot of theoretical distribution. Diameter vs. number of drops. (b) Same distribution, diameter vs. volume percent. (c) Same distribution, diameter vs. cumulative volume.

drop will have assumed a spherical form. Now, if a force is applied to the drop for a short time, it will tend to oscillate between an oblate and a prolate spheroid. If the minor radii becomes too small when the drop is in the prolate (sausage or capsule) form, the drop will "neck down" and break into two smaller drops. This process is pictured in Figure 2.2a,b, and d-f. A process of this type will result in the production of two drops each of about half the volume of the original drop. If the process is more energetic, the drop may break into several drops.

Table 2.1 Volume Data for a Theoretical System

| Diameter (μm) | Volume ($\mu m^3$) | Volume percent | Cumulative percent |
|---|---|---|---|
| 10 | 5235.988 | 0.033 | 0.033 |
| 20 | 41887.902 | 0.264 | 0.298 |
| 30 | 141371.669 | 0.893 | 1.190 |
| 40 | 335103.216 | 2.116 | 3.306 |
| 50 | 654498.469 | 4.1322 | 7.438 |
| 60 | 1130973.355 | 7.140 | 14.579 |
| 70 | 179593.800 | 11.339 | 25.917 |
| 80 | 2680825.731 | 16.926 | 42.843 |
| 90 | 3817035.074 | 24.099 | 66.942 |
| 100 | 5235987.756 | 33.058 | 100 |

If the drop has enough energy to exceed the surface binding forces while in the oblate form, it may move towards a toroidal form and then break into several drops. This process is indicated in Figure 2.2c and g-i. In all cases much smaller satellite drops may form.

If a stream of liquid is injected into another, immiscible liquid with little turbulence, a cylinder may form which is unstable, necks down in several spots, and breaks into drops (Fig 2.3). If the injection rate is such as to produce considerable turbulence, the disruption is faster. Newly formed drops may collide thus forming more smaller drops, some of which may recoalesce (Fig. 2.4). In any of these cases, if the liquid impinges against a surface--such as a plate, blade, or knife edge--action similar to the oblate case will occur, and many small drops will be formed.

In actual cases several of these processes are going on simultaneously. The details of any particular dispersion process are also affected by the viscosity of each phase, the shear in the system, the interfacial energy, and the presence of solid particles or dissolved substances.

Each of these dispersion mechanisms tends to produce a typical particle size distribution. Slow leakage of oil into an aqueous system with low shear will result in large drops. Viscous dispersed

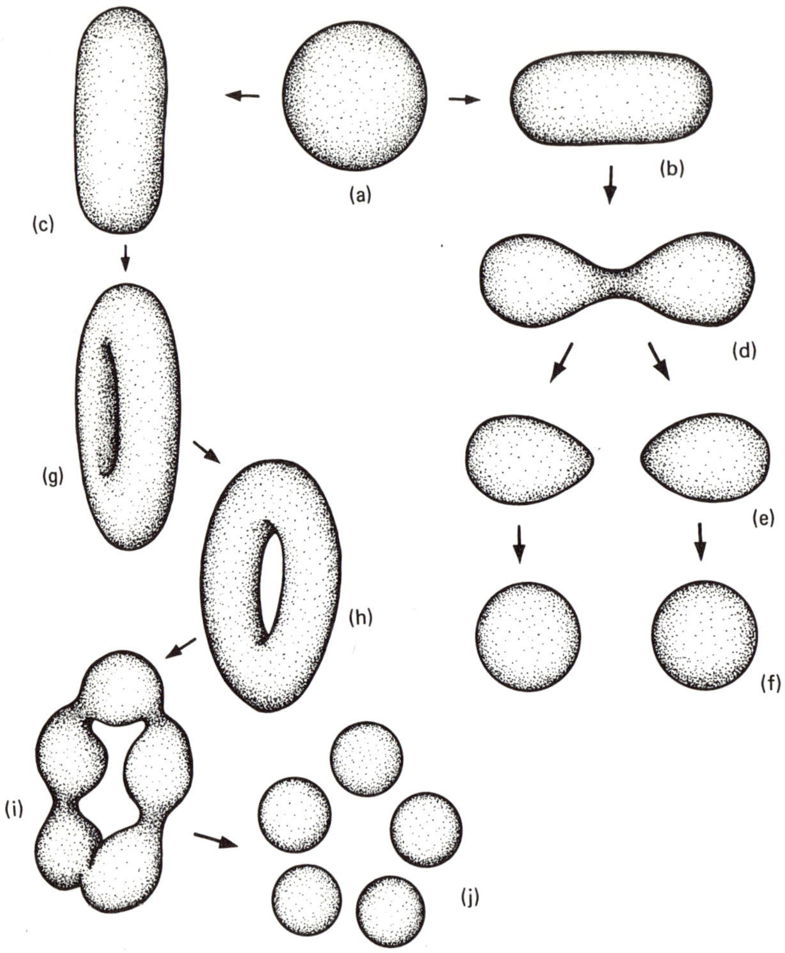


Figure 2.2 Drop breakup mechanisms.

phases tend to produce larger drops. Drops detaching from a tip tend to be nearly the same size and to produce much smaller satellite drops. The presence of shear produces smaller drops and broadens the distribution. The longer the shear is applied, the more distribution shifts to the smaller sizes, but only up to a point. The production of particle sizes in the submicron range requires

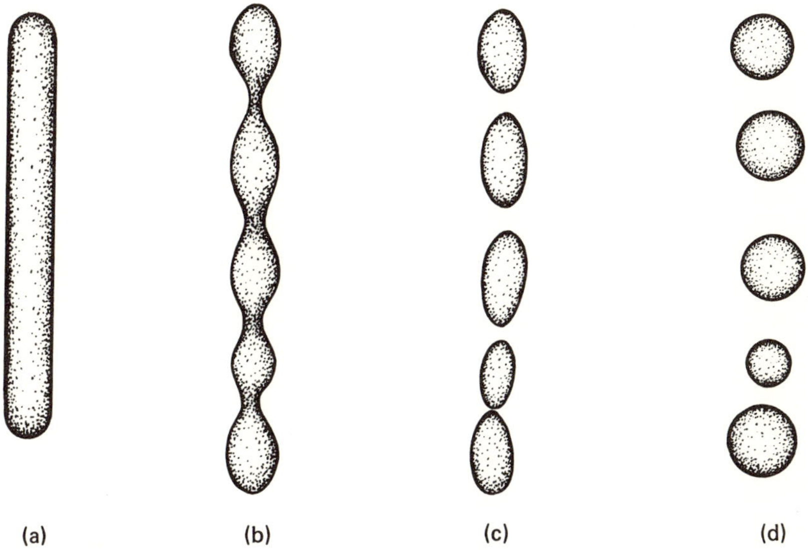


Figure 2.3 Breakup of a stream of liquid, low turbulence.

special conditions. Submicron particles may be obtained by self-emulsification (spontaneous emulsification), by condensation techniques, or by very high shear such as ultrasonic agitation. This means that if you are trying to study a demulsification problem and you can get a particle size distribution on a typical sample, you may be able to deduce the ways in which the problem was caused. If the particle size is less than 10μm and the distribution is narrow, you may suspect that the emulsification agent was present in the internal phase when it was added, or that it was formed in situ. If the distribution is broad with many large-sized droplets, you may have an easily resolvable emulsion that is being stabilized by excessive agitation. Polymodal distributions may indicate that the contamination has several sources. Changes in the particle size distribution with time may indicate the inherent stability of the system or the efficacy of a particular treatment.

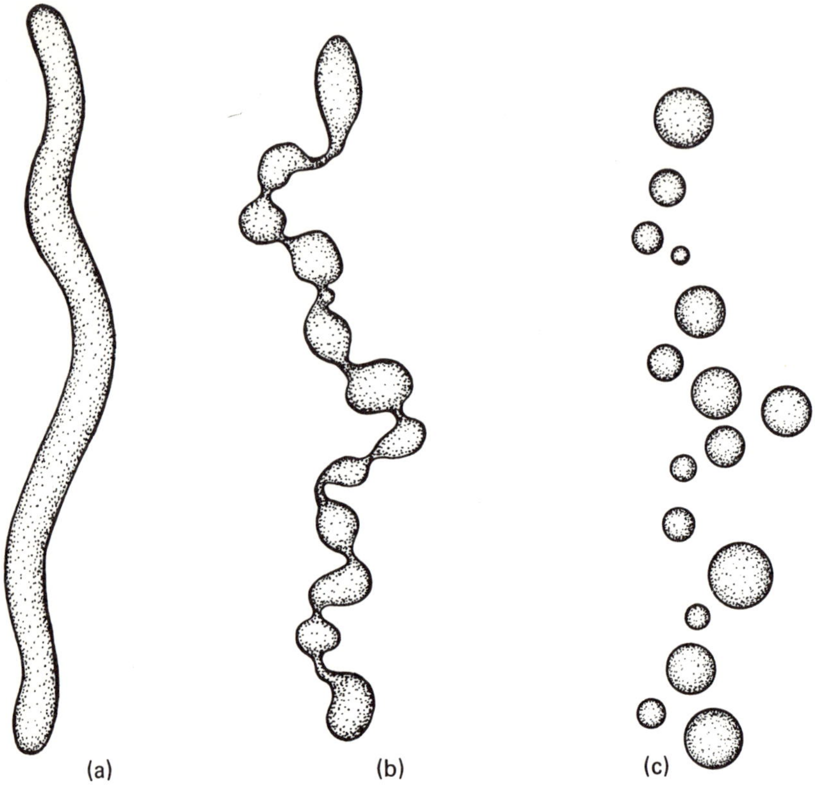


Figure 2.4 Breakup of a stream of turbulent flow.

## PARTICLE CHARGE TESTING

Another property of emulsions that may be measured is the particle charge. The drops may acquire charges in several ways. If ionic emulsifers are present, they can adsorb on the particle and contribute a charge. If the external phase is nonconductive, the drops may acquire a static charge. Charges can be induced on the drops by external magnetic or electrical fields. For oil-in-water emulsions the charge on the particle is an excellent clue to the emulsifier type. We previously mentioned the use of a particle charge tester to test for emulsion type [5]. This device was developed for use in the asphalt-paving industry. Originally, most asphalt emulsions were made by using anionic emulsifiers such as ammonium oleate or sodium oleate. These emulsifiers impart a negative charge to the asphalt particles. In more recent times cationic emulsifiers have come into vogue. Immersing two electrodes in the emulsion and passing a low voltage direct current through the emulsion will cause drops to be attracted to either the cathode or the anode. Thus, the particle charge can be determined. Particle charge, of course, gives clues to the emulsifying or stabilizing agents involved. The use of the particle charge tester in its present form is limited to emulsions with conductive external phases.

## FLOCCULATION AND CREAMING, GENERAL

One way to separate small amounts of oil from an aqueous system is to introduce small bubbles of a gas which will attach themselves to the oil drops and float them to the top of the system. Thus, the gas bubbles act as scavengers for the oil and conduct it to the top where it can be coalesced and skimmed. The bubbles will also tend to float any solids. This fact must be taken into account in any separation process. There are two general ways of introducing the gas bubbles into the system. In one method the gas is simply injected into the water stream through orifices that will produce fine bubbles. In the other method the water is saturated with the gas under pressure. When the pressure is released, the gas comes out as fine bubbles in much the same way bubbles come out of a carbonated beverage when the bottle is opened. Both methods have their staunch advocates. The choice of method probably depends on the specific application. Usually the gas used is air, but methane and natural gas are also used. In cases where oxidation is to be avoided, an inert gas such as nitrogen can be used. Most oily materials will spread over the surface of a gas bubble and thus be trapped and carried to the surface. If the methane or natural gas is used as the floatation gas, measures must be taken to avoid ignition or explosion. Laboratory testing of floatation methods and flocculation processes is not easy. There are some simple tests that can be run, but most experts prefer to follow these tests with pilot plant or small-scale plant tests.

## JAR TESTS

One of the oldest tests in this area is the "jar test". It was originally established, and is still used mostly, for testing flocculants in the removal of solids. It can be adapted for emulsion testing. The equipment usually consists of a multiple stirring unit capable of stirring six jars at a time. It is set up so that each is stirred at the same speed. The rate of stirring is adjustable from about 20 rpm to 120 rpm. The basic procedure is given in ASTM D2035-74 Rev. 1978 [37]. The stirrer unit can be purchased from several laboratory supply houses [38-40]. It was once called a Phipps-Bird stirrer.* The general procedure calls for placing 1000-ml samples of the water to be tested in each of six beakers or wide-mouthed jars. The jars are put in the unit and stirred at about 100 rpm for about 5 min. This is called the fast mix cycle. The stirring speed is then cut to 20-50 rpm and a slow stir, or floc formation, cycle of about 20 min is carried out. The stirring is then stopped, and the samples allowed to settle (or separate) for 5 min to several hours. Any additives or floc aids are added in the fast mix cycle. The duration of each cycle is varied to best model the situation in the real system. If the system being studied is already being treated, the same treatment can be used in the jar test. The different cycles can be varied until results comparable to the actual treatment have been secured. The advantages of different treatments or procedures can then be screened on the jar test.

It must be realized that jar tests do not really represent the same treatment that is afforded in large-scale equipment. Considerable experience is necessary to enable an operator to translate jar tests to field results. However, jar tests are very useful for initial screening and initial assessment of any particular problem. In any particular set of tests, after the settling period has ended, some method must be used to judge the effectiveness of the test. Visual examination will give rough clues. Turbidity measurements can be used but might be misleading. Reddy and Fogler give an excellent mathematical treatment of these problems [41]. If most of the oil is in the form of large droplets with a considerable number of quite small drops, it is possible to remove most of the oil without decreasing the turbidity a like amount. Also, the presence of finely divided solids can mask the removal of most of the oil. The preferred method of the assessment is to withdraw an aliquot of the settled water and extract it with a solvent such as Freon 113 (trichlorotriflouroethane). The concentration of oil in the solvent can then be read with an infrared spectrophotometer [42]. This method determines the "total oil" in the sample and eliminates the

---

*Phipps & Bird, P.O. Box 27324, Richmond, VA, 23261

interference of inorganic ions or solids. It must be remembered that the definition of total oil and grease is simply those materials that are soluble in the selected solvent, Freon 113.

Once preliminary jar tests have been run, it is usual to go to bench scale flotation tests. These bench scale tests are best run in a device which closely models the actual field installation. Several manufacturers of full-scale flocculation equipment also make bench scale units. Most of these units were originally designed for mineral separations or solids removal. Their use to study emulsion problems requires that the operator become familiar with how they perform on emulsions and how to interpret the results for full-scale processes. There is no easy way to tell a novice how to do this. However, a reasonable amount of actual hands-on work in a laboratory with material that is being successfully treated in the plant will soon enable one to secure useful answers.

As mentioned before, there are two main flotation processes—dissolved air flotation (DAF) and induced air flotation (IAF). Dissolved air flotation can be of two subtypes: full-stream pressurization or partial-stream pressurization. As the term implies, in the first type the full stream to be treated is pressurized and saturated with the flotation gas. The full stream is then conducted to the flocculation chamber, and as the pressure is released, the gas forms tiny bubbles which can sweep out the oil. In this method the flocculation aids and other treatment chemicals are added to the full stream just before or in the aeration chamber. Partial-stream pressurization is often used in some systems where the floc that is formed is easily broken or redispersed. A portion of the clean effluent is diverted to a pressurization chamber and saturated with gas. The incoming emulsion stream is treated with the appropriate chemicals and mixed thoroughly but not necessarily violently. This stream is then admitted to the flocculation chamber where the saturated side stream is admixed. The gas can then do its work in a relatively low-shear environment.

## BENCH SCALE TESTS

The bench scale apparatus for conducting dissolved air flotation tests consists of a chamber for saturating the fluid with gas and a flocculation chamber. These chambers can be of several, somewhat different designs. A description of such a test is given by Barcus in Chapter 4 of *Industrial Oily Waste Control* [43]. A typical setup found to be useful in our laboratories is shown in

Figure 2.5.* The pressurization vessel is made from cylindrical polymethylmethacrylate tubing, 4 in. in diameter, with 0.25 in-thick walls. The top and bottom of the unit are 5-in squares of 1-in-thick Lucite or Plexiglas into which a groove has been milled to take the ends of the tube. The whole unit is assembled with gaskets and four 5/16-in stainless steel allthread rods. The top has three 1/4-in female pipe thread (FPT) openings–one for chemical addition, one for the pressure gauge, and one for a safety pressure-relief valve. The bottom also has three 1/4-in. FPT

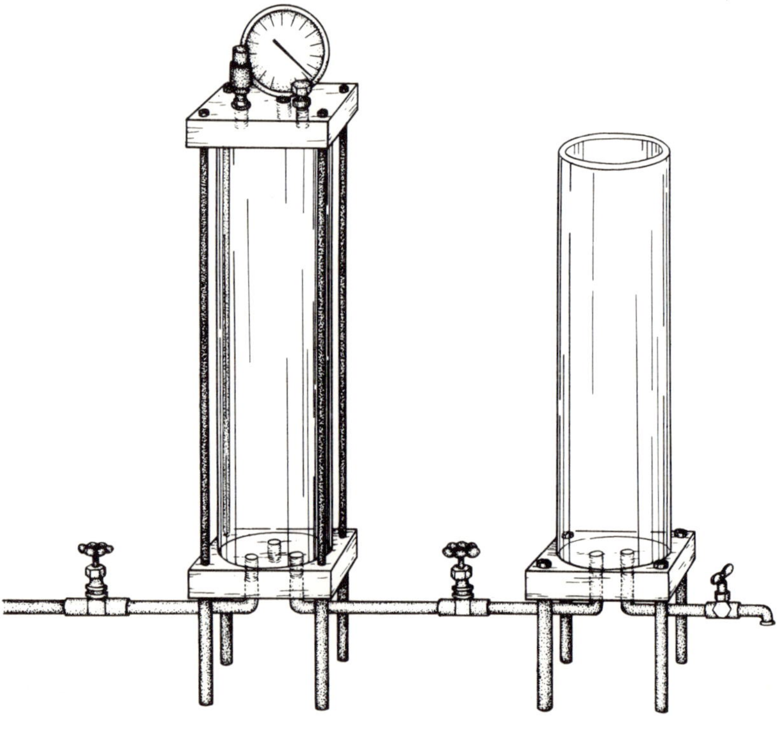

Figure 2.5 Bench scale, dissolved air flotation unit.

*This unit is custom-made by Mr. Fred Jenkins, 10307 Richview Dr., St. Louis, MO 63127. Telephone 314-822-3152

openings—an inlet, an outlet, and a gas supply opening. The flocculation chamber is also made from polymethylmethacylate. The cylindrical body is solvent welded to the square base. The base has two 1/4-in. FPT openings—one for introduction of the emulsion and one for removing clarified samples. The body of each unit is 14 in. deep. Each unit stands on four 4-in. by 0.5-in. in diameter rods as legs.

To conduct a full-pressurization test the sample to be tested (about 1 liter) is measured into the pressure chamber, chemicals are added, and the chamber is hand-shaken to mix the contents. Gas or air is then introduced through the bottom gas inlet until it begins to escape from the pressure-relief valve. A slow stream is bubbled through the unit for a specified time. The gas is shut off, and the unit is connected to the flocculation chamber with flexible tubing. The test sample is allowed to flow into the flocculation chamber through the needle valve on the outlet at a predetermined rate. The sample is then observed for separation, and samples may be drawn from the bottom at intervals for residual-oil determinations. Oil may be skimmed from the top for analysis.

The partial-stream pressurization tests are run in much the same manner except that most or all of the test sample is put into the flocculation chamber, and the chemicals are mixed gently. Then either a smaller portion of the emulsion or a portion of the clarified effluent is pressurized in the pressure unit and introduced into the flocculation chamber. The test then proceeds as before. Observations are made at regular intervals to determine the size and amount of bubbles released and the amount of oil removed.

Induced air flotation is also tested in bench scale equipment. Two typical bench scale units will be described. The first uses a rotor and stator to form a vortex and draw in gas.* Such a unit is pictured in Figure 2.6. The unit is on a base about 22 in. long and stands about 22 in. high. It consists of a specially molded Pyrex glass flotation cell, an air induction unit, and a motor to drive the unit. The motor has a speed control and a rpm indicator. The air induction unit consists of a hollow, cast metal stator-housing about 10 in. long with two rows of holes around the periphery of the bottom. A drive shaft passes through

*One model of this type is the Wemco Lab Flotation Machine, Manufactured by Wemco Div. of Envirotech Corp., 721 N. 8th St., Sacramento, CA

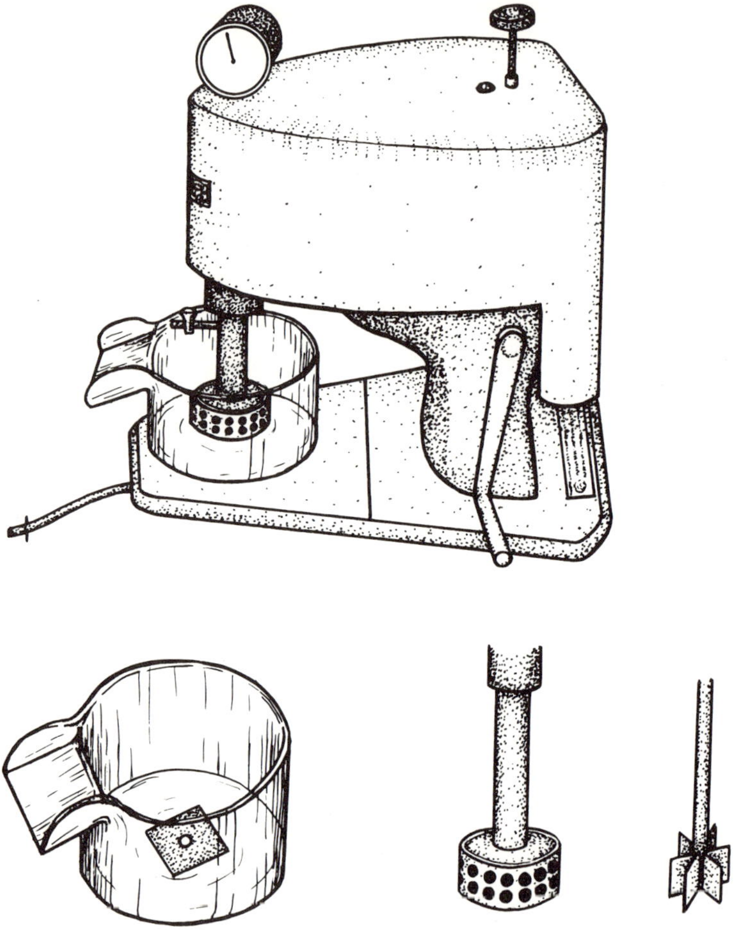

Figure 2.6 Induced air flotation unit by Wemco.

the hollow interior and terminates in a six-bladed impeller molded from an elastomer. There is a port in the outer housing to allow air to enter. The flow of air is controlled by a petcock. In operation the motor drives the impeller which forces fluid out the bottom of the unit and through the holes in the stator. This draws air through the port and mixes it with the water. In passing by the impeller and through the holes, it is broken into fine bubbles which rise through the test sample in the flocculation cell. The cell has a molded lip from which the entrained oil can be skimmed with a skimming tool (a $2\frac{1}{2}$-in. paint brish works well). There is a port in the bottom of the flotation cell for the removal of test samples.

In conducting the test a 3-liter sample of test emulsion is placed in the bowl. The air inlet valve should be closed. The motor is started and the desired speed adjustment made. A typical setting is 900 rpm. Treating chemicals are added to the bowl and mixed for 15 to 30 sec. The air inlet is now opened, and aeration is allowed to proceed for a set time, typically 2 min. During this time the oil and other waste should be skimmed from the top. Do not overskim. The unit is now stopped, the sample line flushed, and a sample drawn form the bottom of the bowl. This sample is analyzed for oil and grease by standard methods.

The other induced air flotation bench test apparatus is known as the nozzle air unit. It is pictured in Figure 2-7. It consists of a rectangular flotation cell constructed from polymethylmethacrylate. The unit is about 8 in. on a side and about 6 in. deep. It has a fluid outlet and a sample port near the bottom on opposite

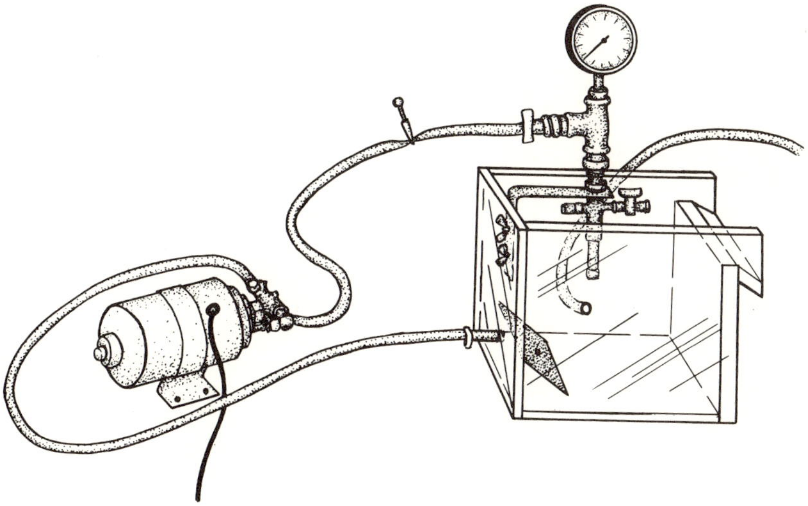

Figure 2.7 Induced air flotation unit, bench scale.

sides. A small centrifugal pump is connected to the fluid-outlet port with flexible tubing. A small plate is fastened inside the cell over the outlet port to reduce the turbulence caused by withdrawing fluid. The outlet of the pump is connected to an aspirator unit which has an ordinary venturi tube to draw in air. A pressure gauge in the line measures the pressure fed to the venturi tube.

Tests are run by filling the chamber to just below the overflow wire. Recirculation is begun and the pressure on the gauge is adjusted by controlling the speed of the pump or by throttling the flow. Treating chemicals are added and mixed for a short time (15-30 sec). The air-inlet-valve is now opened, and the pressure is adjusted if necessary. Flotation is continued for a set period, typically 4-5 min. Observations are made on the speed of floc formation and the amount and form of the floated material. The top is skimmed as in the previous test. The pump is shut off, and a sample is drawn from the bottom of the unit. Determinations are made of oil and grease and/or turbidity.

## SUMMARY

These are typical tests that can be run on emulsions. Other specific tests will be discussed in later chapters.

## REFERENCES

1. J. Newman, *Phys. Chem. 18*:34 (1914)
2. J. Woodman, *J. Soc. Chem. Ind. 51*:95 T (1932)
3. R. J. Hartman, *Colloid Chemistry*, Houghton Mifflin, New York, 1939, p. 341
4. J. Hall, *Phys. Chem. 21*:616 (1917)
5. *Book of ASTM Standards*, Part 15, D244, ASTM, Philadelphia, 1976, p. 231
6. Orelup, *Progressive Perfumery and Cosmetics, May*:53 (1938)
7. P. Sherman, *Emulsion Science*, Academic Press, New York, 1968, pp. 207-208
8. T. B. Robertson, *Kolloid Z. 7*:7 (1910)
9. *Book of ASTM Standards*, Part 23, D1479, ASTM, Philadelphia, 1980, p. 784
10. Official Methods of Analysis of the Association of Official Analytical Chemists, A.O.A.C., Washington, D. C. 12 ed., 1975, p. 259

11. *Book of ASTM Standards*, Part 31, D2778, ASTM, Philadelphia, 1980, p. 630
12. *Book of ASTM Standards*,Part 31, D2910, ASTM, Philadelphia 1980, p. 625
13. *Book of ASTM Standards*, Part 23, D96, ASTM, Philadelphia, 1978, pp. 63-80
14. *Book of ASTM Standards*, Part 15, S244, ASTM, Philadelphia, 1979, pp. 227-241
15. *Book of ASTM Standards*,Part 23, D95, ASTM, Philadelphia, 1978, pp. 58-62
15a. Lissant, K. J. U.S. Patent 3,378,418 April 16, 1968
16. P. Becher, *J. Colloid Sci. 19*:468-472 (1964)
17. A. King and L. N. Mukherjee, *J. Soc, CHem. Ind. 58*:243
18. J. O. Sibree, *Trans. Faraday Soc. 27*:161 (1931)
19. W. N. Maclay and E. Ml Gindler, *J. Colloid Sci. 18*:342 (1963)
20. F. Endter and H. Gebauer, *Optik 13*:97 (1956)
21. M. Wales, *J. Phys. Chem. 66*:1768 (1962)
22. M. L. Wallach, W. Hellis, and A. F. Stevenson, *J. Chem. Phys. 34*:1796 (1961)
23. M. Kerker, *The Scattering of Light and Other Electromagnetic Radiation*, Academic Press, New York, 1969.
24. S. H. Maron and M. E. Elder, *J. Colloid Sci. 18*:107-118 (1963)
25. S. H. Maron and M. E. Elder, *J. Colloid Sci. 19*:591-601
26. G. Mie, *Am. Physik 25*:377-445 (1908)
27. C. K. Sloan, *J. Phys. Chem. 59*:834 (1955)
28. W. H. Aughey and F. J. Baum, *J. Optical Soc. Am. 44*: 833 (1954)
29. *Instruction Manual for Sedigraph Particle Size Analyzer*, Micrometritecs Instrument Corp., Norcross, Georgia, 1978
30. Coulter Product Information Bulletin 4203962 Rev. 3-75 Coulter Electronics Inc., Hialeah, Florida
31. W. I. Higuchi, presented at Sci. Section of Am. Pharm. Assn., Miami Beach, Florida, May, 1963
32. E. L. Rowe, *J. Pharm. Sci. 54* (1962)
33. R. E. Wachtel and V. K. LaMer, *J. Colloid Sci 17*:534 (1962)
34. M. Schrenzel, *Pharm. Acta, Helv. 41*:218-277 (1966)
35. T. R. Lien and C. R. Phillips, *Envirn. Sci. Tech. 6*: 558-561 (1974)
36. R. Clift, J. R. Grace, and M. E. Weber, *Bubbles, Drops and Particles*, Academic Press, New York, 1978, p. 321 ff.
37. *Book of ASTM Stnadards*, Part 31, D2035 ASTM, Philadelphia, 1978, pp. 1242-1246

38. SGA Scientific Inc., 735 Broad Street, Bloomfield, N.J. 07003, 1976 Catalog p. 985
39. Curtin Matheson Scientific Co., P.O. Box 1564, Houston, Texas, 77--1, 1979-80 Catalog p. 994
40. Sargent Welch, 7300 N. Linder Ave., P.O. Box 1026, Skokie, Illinois, 60007, Catalog #128, p. 1227
41. S. R. Reddy and H. S. Fogler, J. *Colloid Interface Sci.* *79*:101-104 (1981)
42. *Standard Methods for the Examination of Water and Waste Water*, 14th Ed. APHA, AWWA, WPCF Method 502B, P. 516-517
43. W. J. Barcus, in *Industrial Oily Waste Control*, (R. M. Skallerup, ed.) American Petroleum Institute and American Society of Lubrication Engineers, 1971, p. 23-25

# 3

# Demulsification of Oil-in-Water Emulsions

## INTRODUCTION

For most people oil-in-water emulsions are the only type of emulsions known. This conception is incorrect, but even in technical discussions one finds the implicit assumption that when one refers to "emulsion" the external phase is assumed to be aqueous. There is considerable superficial evidence for such an attitude. In our bodies and our world, water is pervasive. Plant and animal fluids are almost invariably water based. The strange thing is that although we usually consider emulsions to be water external, we don't seem to stop to think how common oil-in-water (O/W) emulsions are. The first person who churned butter was breaking an oil-in-water emulsion and inverting it. The natural plant saps from which we get rubber and the juices we press from olives, coconuts, and other plant parts are often dispersions of oily materials in water. We concentrate them until they invert, and we have to deal with "wet oils". The washing process involves the formation of a dispersion of oily wastes in the wash water. Oils, lubricants and fuels are spilled, dripped, or "disposed of" and then washed away by rain or deliberate flushing.

It is important to remember that every component of a system and everything that is added to the system to treat it must eventually be accounted for. If you have to clean up an effluent stream, you will have to dispose of whatever you remove. This means that as you plan your separation strategy, you must take into account the form in which you will leave each fraction.

As an example, you may be able to remove a small amount of oil from a water stream by passing the water through a clay filter that absorbs the oil, but you will have to clean and repack the filter from time to time, and the disposal of oily clay will have to be included in your plans. If it is to be used as landfill, you must be sure that it will not come back to haunt you later. The job of demulsification is not really through until you have found a place for each fraction in the system.

One of the main reasons for the recent interest in demulsification of O/W systems is that the above principle has been ignored. As soon as something could be flushed down a sewer or into a creek, it was felt that it had disappeared. Many disposal systems still operate on the principle of dilution. If the limit on the concentration of a particular material is 5 ppm, the waste stream is diluted until this specification is met. Ultimately, this is a fools' game. If the material really should be kept out of the biosphere, it should be kept out, not spread around. There is a real, practical reason for this. It is not some obscure moral principle. Even if recycling is not planned at the present time, stockpiling the waste in a form that will allow later reclamation should be considered. We are already leaching old copper spoil piles and other mine dumps.

Consequently, we have to plan not only to separate the oil from the water, but also to consider the form in which to recover it. Can it be fed back into the process? Can it be used as a fuel? Can it be sold for some special purpose? Can it be stored and batch reclaimed from time to time?

## OCCURRENCE

Where do oil-in-water emulsions which must be broken occur? A few examples are listed below to illustrate the international scope of the problem. The Russians talk of the purification of concentrated, oily waste waters from machine and tool works [1]. The Germans have a patented apparatus for the purification of cutting-oil emulsions or "grinding waters" [2,3]. The French write about the treatment of waste waters from cold-rolling mills [4]. British Patent 1,192,541 teaches the recovery of the oil content of waste soluble-oil emulsions [5]. The Belgians write of the separation of the oil from waste waters in railroad shops [6]. The Japanese discuss the treatment of waste oil from nonferrous metal processing [7].

The treatment of pickling plant effluent [8], emulsions, aluminum cold-rolling plants [9], and emulsions of waste spinning oils in textile plants are cited as problems [10]. Effluents from citrus-processing plants contain oils which must be recovered [11], and the U.S. Navy encounters emulsion problems in the treatment of bilge water [12]. Clearly, the literature is full of references to O/W emulsion problems. We have not even mentioned the problems in municipal sewage treatment plants or the cleanup of marine or land oil spills. Nor have we considered the runoff from the feed lots or agricultural lands. From the numerous examples cited it is evident that unwanted oil-in-water emulsions are common and the need for breaking them is increasing.

The laboratory and field tests that can be applied to emulsions have already been discussed. It is appropriate to consider full-scale operations. The literature provides us with many examples. Dressel reports that it was accidentally discovered that aqueous drilling-oil emulsions could be broken by the addition of "soda" [13]. Hazeltine [14] reports tests on several waste, soluble oil, emulsions made from specified commercial products. He finds that each product requires a somewhat different optimum treatment, but that the addition of "alum and lime" is, in general, most effective. A series of tests by Pincus, Ockert, and Kinney [15] on the emulsions formed in the manufacture of enriched water gas fails to find any obvious relationships between the stability of the emulsions and the amounts of asphaltenes, free carbon, or uncracked oil they contain. The use of heat does not resolve the problem. The relative effectiveness of several surface-active agents is discussed. There is a Russian report that discusses the use of filters to remove oil from waste water [16]. They find that the use of a mixture of hydrophilic and hydrophobic fibers in the filter cloth increases the efficiency of the process. The process has to be carefully controlled to prevent breakthrough of the oil. A detailed discussion of the cleanup of airplane wash-waters in Germany is given by Gummels [17]. Although this report was written in 1958, it is typical of many procedures today. The waste water is treated once a day and for this purpose is collected in a bulk tank. Sand, grass, and other debris are filtered out and any free oil is allowed to separate and is then decanted. The emulsion is said to be broken by electrolysis. The water is then treated with ferrous sulfate and lime and allowed to settle. The clarified water is reused or run to the sewer. It is stated that the clarified water is nontoxic to trout or carp.

A similar report is given of the same problems occurring in the cleaning of U.S. Air Force equipment [18]. It is reported that the addition of calcium chloride will break the emulsion. Air flotation is of little help if enough retention time is available. Filtration through sawdust is a possible alternative. Chemical feed rates must be carefully controlled.

During the processing of olives to recover olive oil, an emulsion results when the olive paste is diluted with water. Castorina reports that this emulsion can be broken by subjecting it to electrophoresis [19]. Balden outlines the methods used by the Chrysler Corporation plant in Detroit, Michigan, to treat its wastes. A pilot plant that uses flue gases to lower the pH of an oily waste stream and so break the emulsion is described. He gives detailed flow-diagrams [20]. Farr et al. describes their solution of a production problem in an extractor using tests in a lab scale unit. They find, in the lab, that by pumping the ether layer and the extractor bottom in the proper ratio during recycling, they break the emulsion. This method also solves the problem in the plant equipment [21].

Douglas states that citrus-oil emulsions can be broken by the addition of electrolylic salts [22]. This is a case of wanting to recover a valuable, dispersed product. We could go on for pages in this way. What patterns begin to emerge? One of the most striking observations is that most books on the subject of effluent cleanup almost ignore the removal of oil, or they relegate it to sludge cleanup. There are several reasons for this. Historically, the main reason for effluent cleanup has been to meet objections by those on whom the effluent was being dumped. Inhabitants of the surrounding areas would claim that their drinking water tasted strange or was colored. They would object to odors. The effluent would kill their food fish or cattle. As a result of this, the efforts of water treatment engineers have been concentrated on eliminating odor, color, or taste then diluting the stream to the point where nobody would complain.

Even today a large proportion of the sewage treatment plants are "primary treatment" plants. This means, essentially, that they strain out the big lumps. Actually, a farm septic tank and drainfield give better treatment to sewage than many municipal plants. But today has brought new problems. For centuries we have used our fields and rivers as our treatment plants, and they served well as long as we didn't overload them. Now we are faced with a dual problem. We are exceeding the natural treatment capacity of the biosphere, and we are running short of materials and water. This puts a different light on the problem. We are beginning to learn that some of the more serious waste-treatment problems are not the most obvious ones.

## TYPICAL TREATMENT PLANT

We are also learning that much of the standard equipment and many of the usual procedures are not well chosen to remove oily contaminants. The usual standard effluent treatment is as follows:

1. Remove large pieces in a "trash trap".
2. Skim off any floating oil.
3. Use flotation or filtration to remove most of the fine particles or liquids.
4. Subject stream to bio-oxidation.
5. Filter.
6. Reuse water or flow to surface system.

A flow chart of this general scheme is shown in Figure 3.1. Note that in each step something is removed and must be used or discarded. If we take into account the secondary disposal steps, the diagram becomes much more complex. An example of this is given in Figure 3.2. The ramifications of such a scheme will be considered in some detail. Assume that, as input, the flow from a hypothetical municipal sewer system is used. Such a flow will have an extremely diverse composition. It will have tramp metal and other dense materials. It will contain cellulosic and other fibrous trash, as well as paper, plastic, wood, rubber, etc. There will be petroleum oils and grease, animal and vegetable oils and fats, food scraps, and fecal matter. The water will also have a burden of dissolved inorganic and organic salts, detergents, and other soluble organic compounds. It may contain industrial wastes, metal oxides and sulfides, paint pigments, halogenated materials, phenols and so on. Not only will the composition be diverse and variable, but the delivery rate will also be highly variable. All these factors must be taken into account in designing a treatment plant.

What does all this have to do with demulsification? These conditions provide several situations where unnecessary emulsions can be made and several places where emulsions will have to be broken. None of the steps can be taken without considering the entire system.

To begin we must separate the large and heavy trash and even out the flow. This is usually done in a settling basin coupled with a trash trap and an oil skimmer. At this point the complications

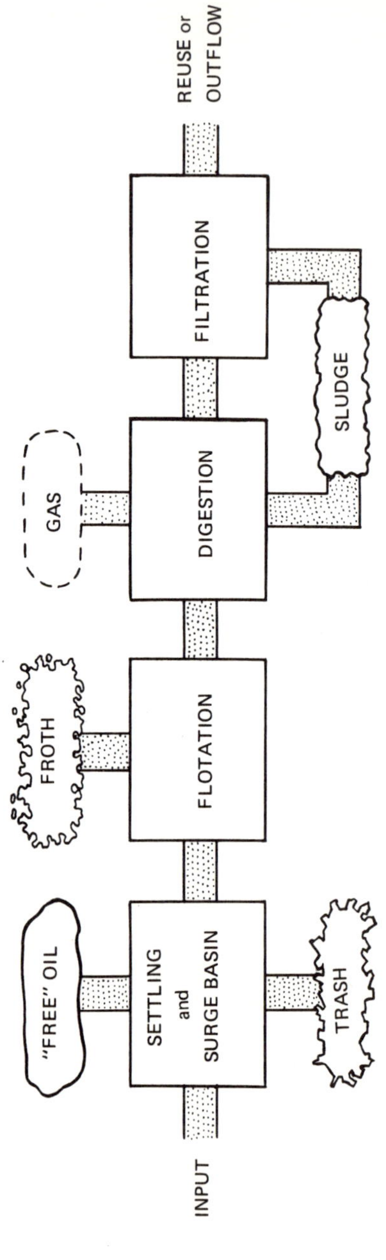


Figure 3.1 General water-treatment scheme.

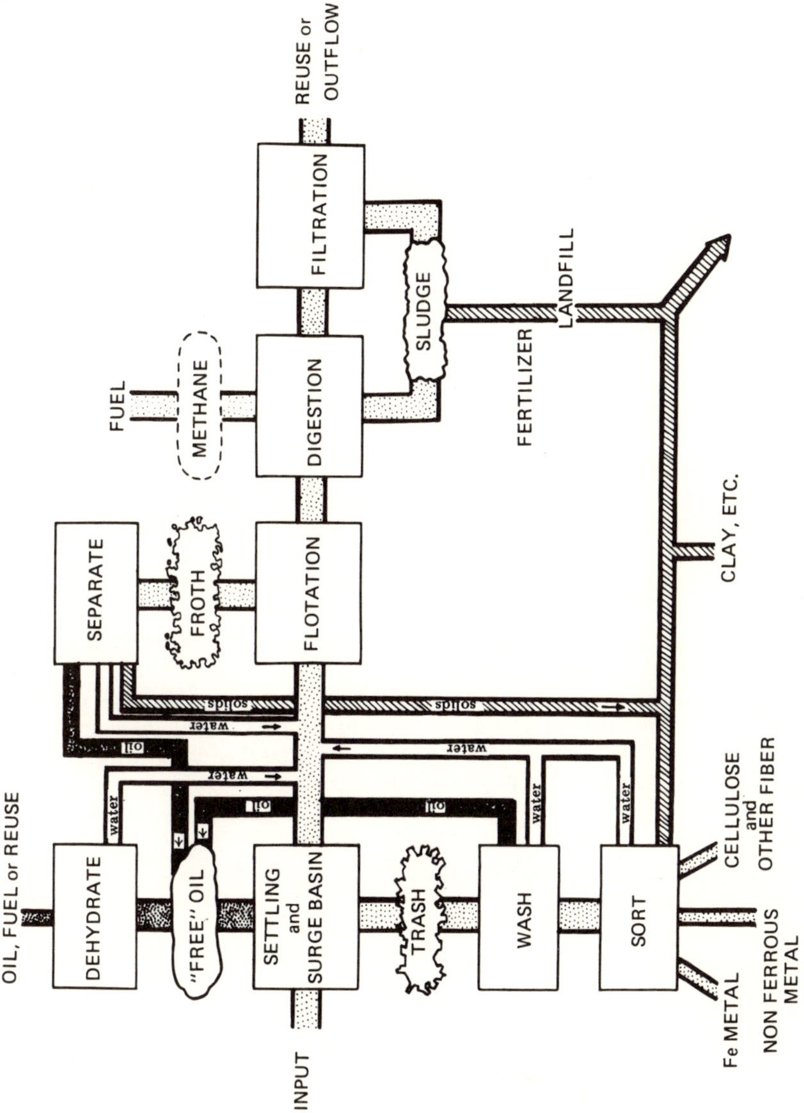


Figure 3.2 Detailed water-treatment scheme.

begin. If there is much bulky trash, it will entrain oil and grease, and at the same time, the oil may wet and entrain small or light solids. In some cases provisions must be made to chop and slurry the trash solids and then separate entrained oil and combine it with the "free-oil" skimmings. It may be necessary to adjust the pH or add demulsifiers to get the oil to separate. The oil-free slurry can then be further separated into metal, paper, pulp, clays, etc.

Throughout this first step the differences in density and the relative wettability of the various components must be considered. Some solids are preferentially oil wetted. Metal sulfides, for instance, are often easily oil wet. Since they are also quite dense, they tend to pull oil down with them. On the other hand, wood fibers and other cellulosics tend to be preferentially water wet and to be lighter than water or about the same density. They therefore tend to rise and carry water into the top oil layer. Sometimes interposing screens at appropriate points in a system can prevent some of these troubles. By now it should be obvious that the best way to solve some of these problems is to prevent them from occurring. Wherever possible, avoid introducing oily wastes into an aqueous stream. Keep trash solids out of sewers. Try to retain solids close to their point of origin and recover them there. In spite of all your planning, however, unfortunate mixtures are going to occur. A traffic accident may dump oil into a sewer. A fire in a manufacturing plant may cause a tank rupture. Water used to fight a fire may wash debris into the drains. All waste-treatment systems should plan for the occasional overwhelming accident.

After the gross trash and floating oil have been removed, we still have to remove suspended solids and emulsified oil. At this point the amount of suspended material and dissolved material may dictate the best course to follow. For ultimate cleanup we will probably use some kind of biological digestion process. If the total biological oxygen demand (BOD) and the chemical oxygen demand (COD) are low enough, it may be possible to leave out a filtration or flotation step. In general, however, biological processes do not work well on streams containing much oil.

The decision as to whether to go to filters or flotation will also depend on the average composition of the stream. Filters may give trouble if the suspended material contains much oil. If oil is predominant, flotation may be the best choice. In either case plans must be made to deal with the material that is removed. The filter cake, the "froth", and the backwash stream, should be routed to the appropriate oil-treatment or solid-treatment unit.

The use of flotation units at this stage in a treatment system is becoming more widespread. They have the advantage of being able to handle both solids and oily wastes. In the past it has been standard practice to add iron and alum to provide a floc. This method is quite effective for solids removal but increases the bulk of the froth to be handled. The recent trend is to minimize the use of inorganic flocs and depend largely on organic polymers. If much oil is present, a demulsifier may also be used. After a well-designed filtration or flotation step has been completed, main flow can usually be dealt with in a conventional activated-sludge treatment plant or other biological system.

Now let us go back and study in detail how to deal specifically with the separation of oil from aqueous streams. We have seen how the need to do this is often ignored in present-day systems. We will now investigate in some detail how to isolate and treat oil-in-water emulsions wherever they are encountered. ("Reverse" petroleum emulsions will be dealt with elsewhere). We will look at several pieces of typical commercial equipment and see how the general principles apply to each. In doing this we will select certain specific units for examples. It must be remembered that a variety of commercial units are available and not all can be mentioned. The selection of the equipment of one particular manufacturer as an example does not imply that that unit is the best or the only one available.

## CYCLONE SEPARATORS

The oil separation and some of the solids classification can sometimes be done at this point with a device known as a cyclone separator. Cyclone separators are widely used in the paper industry where they are sold under several trade names including Centri-cleaner, Hydrocyclone, Assablador. They like a centrifuge, depend on centrifugal force to speed separation. They differ in that they have no moving parts. Figure 3.3 shows the general form of a cyclone separator. The unit consists of a conical shell of metal or plastic with an inlet set tangentially at the large end. The liquid (or slurry) is introduced under pressure and is forced by the shape of the unit to spin at a high speed. The liquid spirals down the cone, and then rises up the center around a vortex, and leaves through an outlet at the top. Heavy particles are forced against the sides of the cone and exit at the bottom. Separation is on the basis of differences in density. The selectivity of the separation can be varied by the size of the unit, the pressure drop, and the size of the reject orifice. The shape of the particles also has an

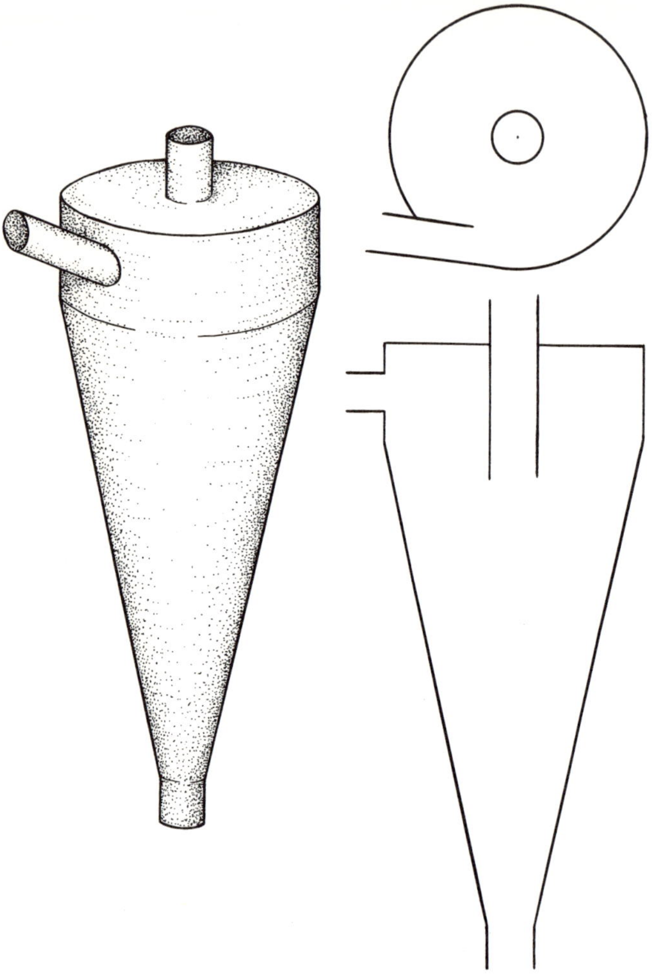

Figure 3.3 Cyclone separator.

effect. Cyclones are quite effective in removing dense, chunky particles without removing fibers and other material with a density near that of water. The throughput rate of cyclones is relatively low. They are usually used in parallel banks or in stages.

## CENTRIFUGES

We have not mentioned centrifuges as a means of separation up to this point. In general, centrifuges are not as widely used for demulsification as some other methods. Fundamentally, they are not different in principle from gravity settlers or cyclones. They do, however, represent a considerable investment, and they require careful operation, maintenance, and upkeep. They are justified where there is a relatively low-volume, high-value material to recover or where rapid separation is necessary. They also have applications where the fluids to be separated are viscous. They are, therefore, used more often in dealing with water-in-oil emulsions. Centrifuges have been in use for a long time, and procedures for their design and use are well known. Like other settlers they will not of themselves break a stable emulsion. Used with chemical demulsifiers they are effective in a variety of commercial applications.

## "STILLING" PONDS

Next we will look at settling, or "stilling", ponds. These ponds may be simple holes in the ground if the ground is impervious enough to hold the water without seepage. If it is not, they may be lined with plastic, asphalt, or concrete. A settling pond should be at least large enough to hold all the water that is likely to be accumulated in a period of 24-48 hr. This will allow shutdown of the treatment plant for several hours for repairs. It should be remembered that solids will accumulate in the settling pond and will have to be removed with a drag line or other means from time to time.

For this reason it may be preferable to have two ponds, each large enough to function while the other is being cleaned. If the pond has a plastic liner, it will have to be drained and flushed so as not to damage the liner. Free oil will rise to the top of the pond, and volatile material will be lost. Also, the oil layer may become oxidized. In any event, means must be provided to skim the oil and then process it further. Thus, it can be seen that settling ponds may not really be the simplest answer if the suspended solids load is high or if there is much oily material. In such cases a separator may be preferable to a pond. Holding tanks may be used, or a separator may be used before the pond. The use of the pond as a surge basin may be the best procedure.

## TYPES OF SEPARATORS

Separators come in several types and many sizes. It should be remembered that mechanical settling will not remove emulsified oil by itself. However, appropriate treatment of the flow may break the emulsion and make separation possible. The three most common separator types are the American Petroleum Institute (API) separator, the circular separator, and the parallel-plate separator. Each of these separators can be designed to remove all droplets larger than 0.015 cm in diameter. They each have means for removing material that settles to the bottom, and each has means for skimming oil. They are either fitted with trash traps or separate trash traps are installed upstream. Rough diagrams of the three types are given in Figures 3.4 to 3.6. The API separator is a rectangular trough 3-10 ft wide, and 50-150 ft long. It is the standard for many industries. The circular separator has the advantage of taking up less space. The parallel-plate separator can be even more compact.

We spoke earlier of tests that could be run on effluent streams. A simple settling test on a sample of the influent will indicate whether a reasonable separation can be expected with a given piece of equipment. Obtain a truly representative sample of the stream to be treated and mix it well, but do not subject it to more shear than it will encounter in the system. Fill a 1 liter graduated cylinder to the liter line with the fluid and observe what happens over the next 30-60 min. If the retention time of the equipment is known, it can be used as the test time. Note what rises to the top and what settles to the bottom. Note whether the intermediate water layer is clear or turbid. Note how compact the layer on the bottom becomes. Try to get some idea of the rate of the rise of the oil droplets and the settling rate of the solids. Determine the volume of the sludge layer and of the oil layer. Now it may pay to repeat thc test in a 1-liter separatory funnel. This procedure will let you draw off samples of the settled sludge and then of the bulk water phase. Samples can aso be removed from the graduated cylinder using a syringe and a long needle. The tip of the needle should be ground off square. Needles, as supplied, usually have a tapered tip. Now one can analyze the various layers for "oil and grease", total solids, dissolved solids, etc., and predict how the stream will behave in a gravity settler. If the water is clear and bright, the solids are compact and oil-free, and the oil layer is free of solids and contains no water, a gravity settler may constitute a satisfactory treatment. In actual practice it is rare that simple gravity-settling is all that is required. Some water

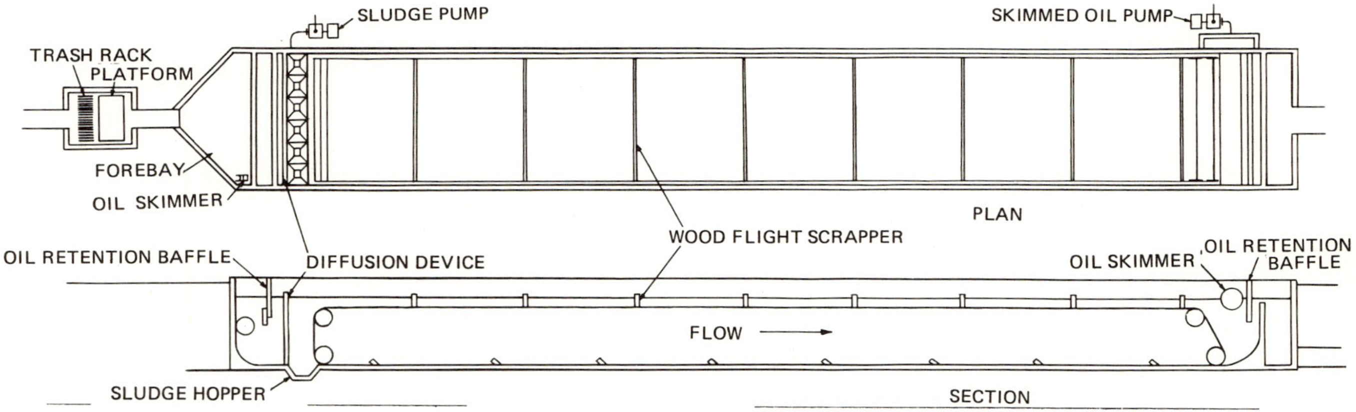


Figure 3.4 API separator.

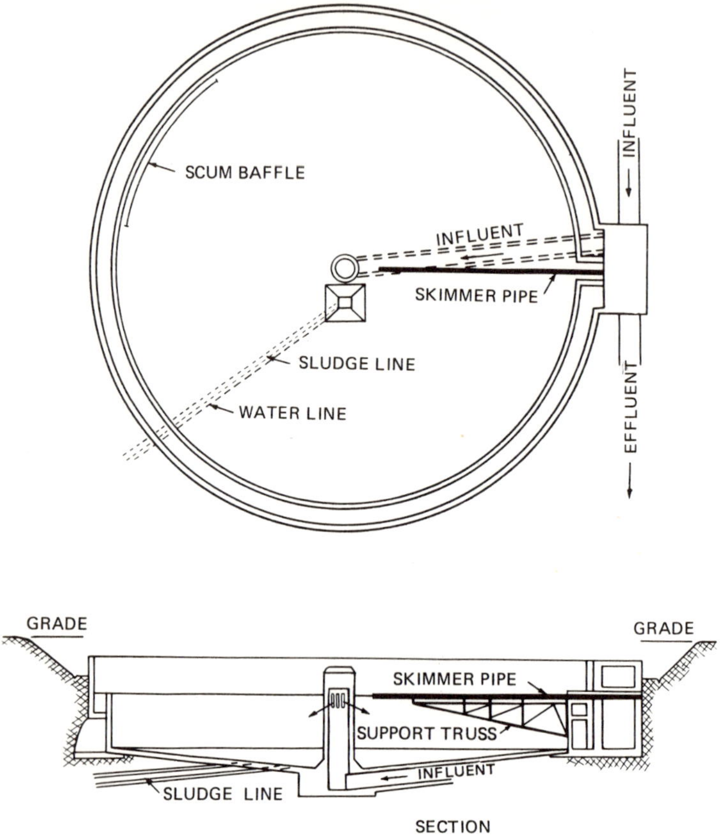


Figure 3.5 Circular separator.

and solids nearly always come over with the oil. Frequently, most of the oil grows down with the solids in the sludge. Before planning the installation of new facilities or changes in old equipment, one should test to see it the treatment of the influent can improve separation. Details of settlers are available in the literature [23-25].

## DESIGN CONSIDERATIONS

Before one begins to design a system, it is necessary to collect information concerning the extremes in the flow rate that can be

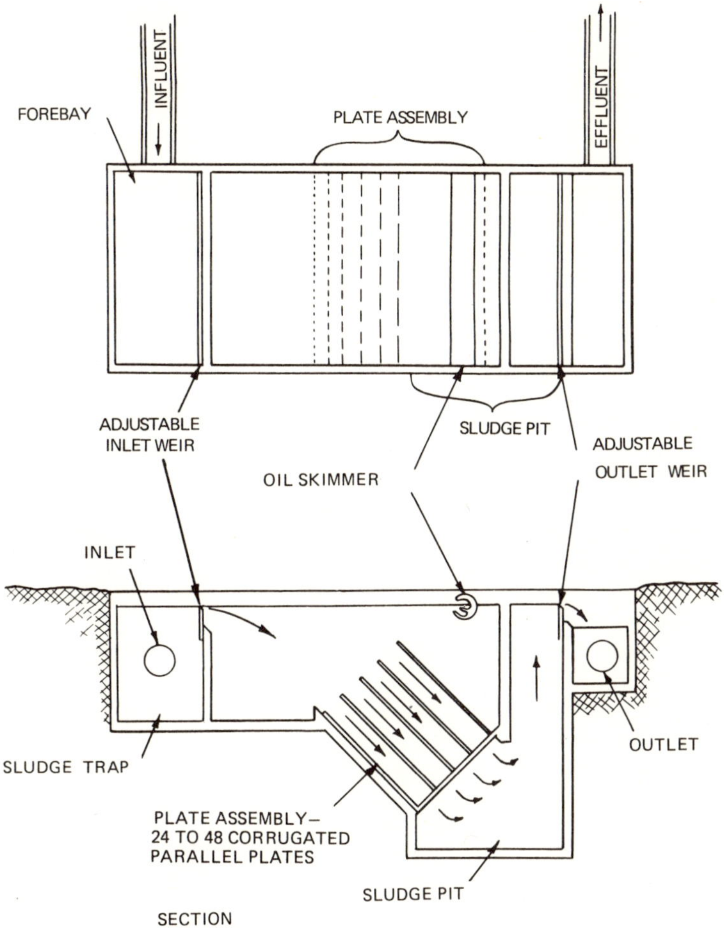


**Figure 3.6** Parallel-plate separator.

expected. One should obtain information disclosing how the flow rate varies from shift to shift and from day to day. Inquiry should be made about conditions on weekends and holidays or during periods of drought or excessive precipitation. Factors which can affect the pH or total dissolved solids (TDS) should be pinpointed and studied. The possibility of isolating certain waste streams which require special treatment should be considered. It is advisable, early in the planning process, to investigate the possibility of combining acidic and alkaline flows to control pH, and the advisability of isolating sanitary wastes.

## FLOCCULATION

So far it has been shown that simple settling tests are a means of predicting the efficacy of the full-scale settling step. Let us consider a water stream with some suspended solids and, perhaps some emulsified oil. If this stream is fairly constant in composition and has a low biological oxygen demand, it may be an acceptable feed for an activated-sludge unit, as is. If not, it will need either filtering or flocculating. The present trend is toward flocculation unless the volume is quite small. Flocculation can be followed by settling again or by flotation. The required flocculants and floc aids are determined by the jar test. This test will also allow testing for methods of breaking the emulsion. Traditionally, flocculation steps have employed inorganic, floc-forming iron or aluminum compounds. The trend now is to avoid using inorganic materials if possible because they add to the volume of the material that must be treated in the sludge disposal step. In many cases the use of some floc-forming compound will be required, but the use of high-molecular-weight polymers and other organic floc aids will greatly reduce the amount of sludge to be handled. Also, if you have a considerable amount of oil to be separated in the flocculation step, the presence of an inorganic floc may complicate matters. Ideally, as much of the solids as possible should be removed in the initial settling step so that when the oil emulsion is broken, the separated oil does not carry solids with it. In practice, this is difficult to accomplish in many streams. There are nearly always some very fine solids, and they may tend to become oil-wet.

After jar tests have allowed initial selection of the optimum conditions for flocculation and indicated what demulsifiers, if any, may help, bench scale flocculation tests, pilot-plant tests, or plant tests will allow finetuning of the treatment. One of the major factors at this point in obtaining trouble-free plant operation is doing all that can be done to assure a constant rate of feed and a constant composition of feed. The treatment and operation of the surge pond or initial settlers should be carefully chosen to smooth out fluctuations in the rate of flow to the flocculators and flotation or settling units. Flocculation processes are particularly sensitive to variations in flow rate and composition.

Once settling and filtration or flocculation have reduced the COD and BOD to reasonable levels, then biological treatment can remove most of the rest of the impurities. Since these processes do not involve demulsification, we do not need to consider them here. We have been discussing the situations where oil-in-water demulsification may be involved. We will now look at some specific examples.

## SPECIFIC EXAMPLES

In the flavoring and perfume fields the active ingredients are often water-insoluble oils that are pressed or extracted from plants. For instance, Douglas [26] patented a process whereby emulsions of citrus oils in water (i.e., citrus pulp) can be recovered by adding solutions of calcium chloride, magnesium chloride, or ferrous sulfate.

In metal-processing plants and machine shops emulsions are used as cooling and lubricating fluids. These emulsions break partially due to the heat generated, and they become contaminated with metal chips and oxidation products. They must either be cleaned and regenerated or separated. A Polish patent [27] discusses the treatment of spent emulsified cutting-oil compositions with chlorinated lime. One liter of spent emulsion is treated with 24 g of chlorinated lime (31% active chlorine) and mixed with air for 10 min. The oil then separates and can be recovered. A Russian article [28] discusses the treatment of waste waters from coil metal paint shops. They use hydrochloric acid to bring the pH to 3 and then separate the oil and remove surfactant by foam fractionation. Schilinski, Frichtmueller and Tscherner [29] and Henley [30] patented the use of sodium sulfate and citric acid as a cleaning and demulsifying fluid in the removal of spent emulsion from metal parts before painting. Beaucaire has patented a process for deoiling waste rolling oil [31] by treating it with waste pickling-acid solution to break the emulsion. The pH is then adjusted to 7-12 with lime, and oxygen is introduced under pressure. The iron present is converted to magnetite which absorbs the oil and is then removed magnetically. Again, the Russians [32] treat waste waters from machine-building plants in a two stage process [1]. Addition of sodium chloride breaks out about 80% of the oil. The water layer is then acidified to pH 4 with sulfuric acid. The rest of the oil separates on the addition of aluminum sulfate. In a chemical engineering symposium Barker and his associates [32] describe how the waste water in a cold-rolling mill is handled. Alum and lime are used to produce an oil floc. Air flotation with periodic skimming separates the oil. It is stated that average operation for 7 months shows removal of 91% of the oil and 81% of the COD. From 2000 to 12,000 gal of oily scum is disposed of each day by using it as road oil. This use may not be allowed today. A German article [33] reports using aluminum sulfate, magnesium chloride, and hydrochloric acid to break emulsions of cooling lubricant oil. Separation is by filtration through sawdust or peat.

During the nitration of toluene for the production of 2,4-dinitrotoluene, an emulsion results. A 1963 article by Somer discusses several means of breaking this emulsion [34]. A Dutch patent application outlines the use of lignosulfonic acid to recover oils from emulsions stabilized by the presence of proteins [35]. Lignosulfonic acid containing sulfuric acid is added to the emulsion. It is heated to 88-90°C for 8 hr. The oil is then separated by centrifugation. The oils recovered by such a process would have to be reusable to justify the long cooking time and the use of a centrifuge.

It should be remembered that milk is an oil-in-water emulsion. When the milk sours, the acid released destabilizes the emulsion, and centrifuges are used to separate out the butterfat. If raw milk is run through a separator, cream is separated. Cream is a concentrated O/W emulsion. If it is acidified and agitated, the emulsion breaks and inverts, producing butter, which is a W/O emulsion. If this emulsion is heated, the viscosity will be reduced, and most of the water will fall out.

In many commercial processes an animal or vegetable product is macerated in water, and a desirable oily material separated from the pulp. It has been found in some cases that the presence of fibrous material prevents or breaks the emulsions that tend to form. Thus, an East German patent [36] disclosed the addition of cellulose pulp to aqueous emulsions of fats to facilitate separation. A U.S. patent [37] covers a similar process where the application of heat and lime in the presence of fibrous materials results in the separation of the oily material which can then be decanted. In another U.S. patent, laundry waste-water is demulsified with lime and some recycled floc [38]. A Japanese patent covers demulsification of spinning-mill wastes by treatment with sodium sulfate [39]. Sometimes the floc-forming materials can be recirculated. A U.S. patent issued to Davidson [40] discloses a process where an oil-in-water waste water emulsion is treated with aluminum sulfate, and the oil is removed by flotation. The floc is then treated with sulfuric acid which dissolves much of the floc and makes the oil easier to recover. The aluminum-rich liquor is then recycled to fresh waste water emulsion.

Besides inorganic compounds, organic polymers can be used to break emulsions. United States Pattent 4,026,794 claims the use of "the water soluble salt of an amphoteric metal and : 10-60:5-40:5-70 acrylamide-diallyl-dimethylammonium chloride-N-vinyl-2-pyrrolidinone coplymer" to treat the water from a diesel servicing facility.

## MECHANICAL COALESCENCE

Mechanical coalescence methods also exist. For many years "hay tanks" have been used to resolve emulsions. Chieu, Gloyna, and Schechter describe the results of a test on fibrous-bed coalescers using various media [41]. Stausser and Kurland have a patent on a method using finely divided, crystalline silica followed by passage through a sand filter [42]. Sedlmeier and Thoenes discuss the treatment of oily steam-condensates [43]. Furness and Allenson have a patent on a process whereby electrodes are immersed in the waste water, and a current causes the generation of small bubbles. These bubbles of gas float the oil to the top of the unit [44]. A Japanese patented device passes the emulsion through a column of coarse quartz sand, and then treats the effluent with lime. Sulfuric acid is used initially to adjust the pH. The precipitated calcium sulfate carries down the remaining oil [45]. A British patent describes the use of a fired board of Saffil fiber. The direction of flow is reversed periodically [46]. Another U.S. Patent discloses an acoustic or shock-wave technique. The O/W emulsion is introduced near the bottom of a tank which is equipped with a steel diaphragm which is made to vibrate at audio frequencies (15-500Hz). Oil is removed form the top of the tank; water is drawn off the bottom [47]. A German patent [48] teaches the passage of the emulsion through a porous electrode such as iron shavings. The electrode supplies ferrous ions which oxidize to ferric ions and cause the emulsion to break. A French article [49] reports the use of a *Technime process*. The details of the process are not disclosed, but it seems to consist of the addition of clay materials with a demulsifier absorbed on the clay. Demulsification followed by flocculation results. The sludge or mud from the process is said to be pelletable. The literature contains many such references [50-53] to mechanical coalescence.

## OVERALL PLANNING

We can now begin to make some generalizations as to how to break oil-in-water emulsions. First, the effect of temperature should be considered. In general, heating large volumes of water is costly, but if waste heat is available, it may be useful. Also a study of the effect of temperature will alert one to the possibility of system malfunction in very warm or very cold weather.

Next the effect of pH should be studied. Often O/W emulsions are sensitive to the presence of acids or bases. It may be possible to break an emulsion using waste acid or a waste caustic from somewhere in the process. It may be possible to avoid the formation of a troublesome emulsion by not allowing a particular acidic or basic waste-stream to mix with the main stream to be treated. In dealing with acidic or basic wastes or in planning pH adjustments it should be remembered that additions will increase the total dissolved solids and that restrictions are being placed on the TDS allowed.

The use of proprietary demulsifiers should also be considered. These products vary widely in composition and usually are blends of several ingredients. Their action is often very specific. For best results the directions and recommendations of the manufacturer should be followed carefully. The selection of the best emulsion destabilizer for a particular case is frequently best done by someone with experience in the particular field.

Having found how to break the emulsion, one must separate the oil from the water. The selection of the specific equipment and procedures is not an easy task. If you know that you will have to deal with significant amounts of oil, it is not wise to leave the design of your system to someone who does not have experience in dealing with oily wastes. Too many waste treatment systems have been installed in good faith without taking into account the differences between systems that deal primarily with solids and those that involve the presence of oils or solvents. Even if a system will usually be dealing with little or no oil, it is well to plan what will be done if some accident results in the introduction of oil. Biological oxidation systems can be severely damaged by excess oil.

The third item to be considered is how to dispose of the materials that are separated from the stream. Solids must be incinerated or buried or recycled. Oils also must be recovered, then reclaimed or burned. It is not wise to depend on some contractor hauling away the oil or solids. If he does not dispose of them properly, you may still be held liable.

Sites for disposal are becoming scarce, and the controls placed on them are becoming tighter. What will you do if you disposal site is shut down? These may not sound like demulsification problems, but they may well dictate the process you use since the oil and solids must be obtained in a form that can be reused or disposed of on a long term basis.

## INVERSION AND PHASE REVERSAL

Oil-in-water emulsions have been considered in this chapter, and water-in-oil emulsions will be covered, in general in the next chapter. We have also discussed in previous chapters the situations which favor emulsion formation. Consider now the conditions that cause emulsion inversion and how some multiple emulsions can form. It will be necessary to consider the geometry of some situations in order to understand what is happening. A general detailed discussion of the geometry of emulsions has been published in several papers and part of a book [54-60]. We will only mention certain major factors here. Consider the simple case of a single drop of oil rising through a water layer, reaching the oil-water interface, and coalescing with the oil layer. This process is shown in Figure 3.7. When the drop reaches the oil-water interface, it rises until it dimples the surface of the oil. Then the film between the drop and the bulk oil phase thins and breaks. The drop then flows into the bulk phase, and any material on the surface of the drop tends to stay at the interface. Thus, insoluble materials tend to accumulate at the interface and any partially soluble, surface-active films also are concentrated there. Eventually this material may become concentrated enough to prevent the further coalescence of the drops. This same process of concentration may be taking place in drops in the bulk waste phase as they rise, touch, and coalesce into larger drops. In any event, a situation may arise where the drops rise to the interface, but do not coalesce. A concentrated layer of drops will form, and as the layer becomes thicker, the drops will be distorted and tend to take on a polyhedral shape. This process is shown in Figure 3.8 This high internal phase ratio (HIPR)

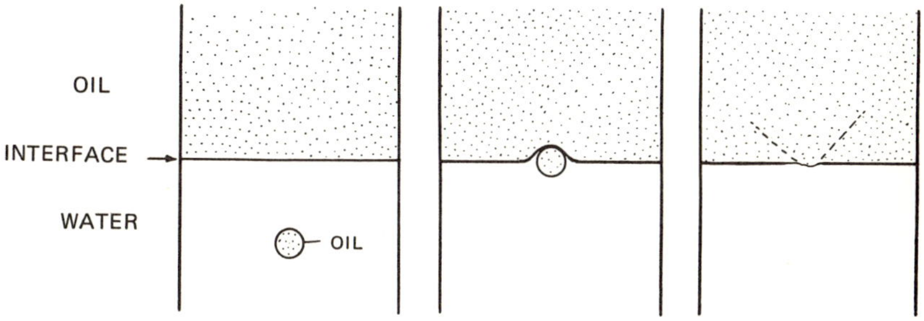


Figure 3.7 Coalescence of a single drop.

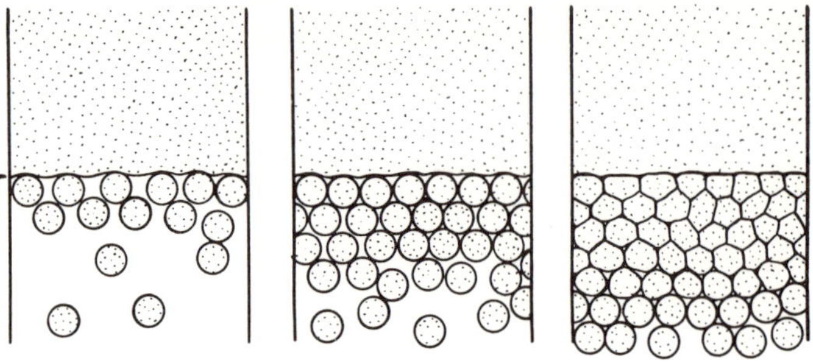


Figure 3.8 Formation of creamed layer.

emulsion will have a much higher apparent viscosity than the oil or the water. It is thick and hard to pump or stir. It is water-external but contains very little water. It will entrap solids. For simplicity's sake consider the drops to be hard spheres and all the same size. When they attain the closest packing, they will occupy about 74% of the total volume and be in a configuration like that pictured in Figure 3.9. It can be seen that there are two kinds of "holes" between the drops: those formed by four drops in a tetrahedral stack and those formed by six drops in an octahedral stack (Fig. 3.10). We will now calculate the relative volumes of these two types of hole. The volume of holes in the tetrahedral stack is equal to the volume of the tetrahedron less the volume of the sections of the sphere included in the tetrahedron. If the radius of the drops is set at unity, then the volume of the tetrahedron is

$$V_T = 0.11785(2r)^3 = 0.9428(r)^3 \tag{1}$$

The volume of the sphere is

$$V_s = \frac{4\pi}{3}(r)^3 \tag{2}$$

The volume of the portion of each sphere included in the tetrahedron is

$$V_p = V_s \frac{\text{(area of spherical triangle)}}{\text{(area of sphere)}} \tag{3}$$

$$V_p = 0.18372\ r^3$$

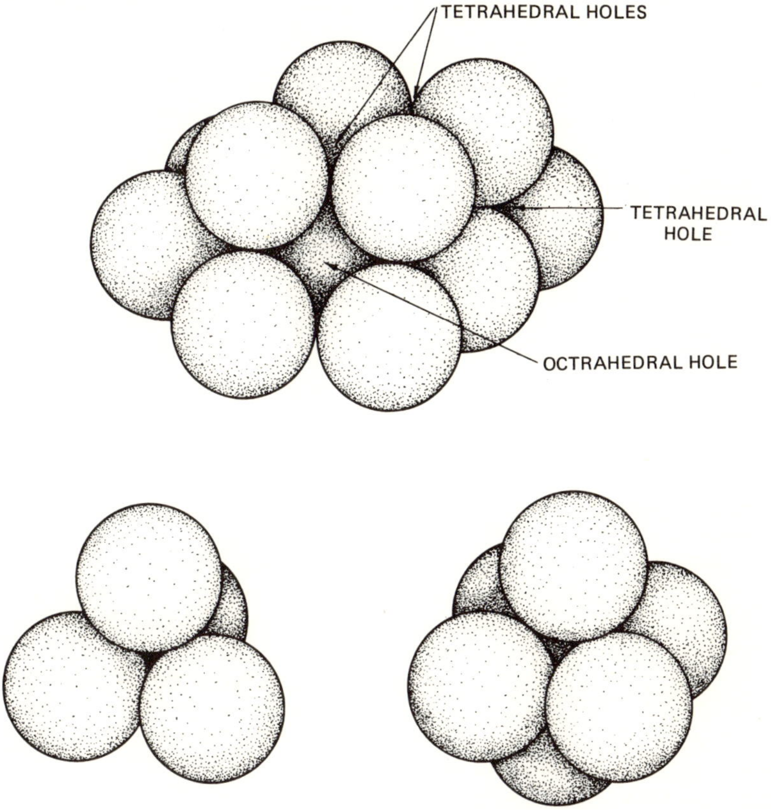


**Figure 3.9** Close packing of drops showing holes.

Thus

Volume of the hole = $0.2079r^3$ or about 5% of the volume of drop (4)

Similar calculations for the octahedral hole yield

Volume of the hole = $V - 1.052\ r$ or about 25% of the volume of the drop (5)

If we assume that the drops of the internal phase flatten as they are crowded, the holes become smaller and more isolated. A rapid

Figure 3.10 Shapes of the holes between drops.

disruption of the interfacial films would then result in an inversion of the emulsion and the production of an emulsion with much smaller droplets. In fact, this technique is used in the commercial production of wax-polish emulsions. Water is added to the hot wax until the water-in-oil emulsion inverts into a very finely divided oil-in-water emulsion which is then cooled rapidly below the melting point of the wax. By similar methods it is possible inadvertently to produce very stable emulsions from the sludge of separation processes. In the use of the settling tanks, the oil is drawn from the top and the water from the bottom while the interface is allowed to build up. When the tank becomes clogged with the interfacial layer, this layer is drawn off and pumped to another tank for further treatment. The high shear to which the sludge layer is subjected in the pump may cause either an increase in emulsion stability or an inversion of the emulsion to a form more difficult to treat.

We are reminded again that emulsions are "constructs". In most instances they do not form spontaneously. It is almost always better to avoid the formation of an emulsion than to try to treat it. Sometimes emulsions cannot be avoided. In all washing operations the use of a detergent to suspend the solids and disperse the oily soils is important to the success of the cleaning process. Emulsions in these cases are almost inevitable. Careful control over the washing procedure or the cleaning operation may be able to save a good deal of trouble.

Whenever detergents or cleaning formulations are added by hand, there is a strong tendency to use too much. In many cases over twice as much material as is needed is used. Tests can be run to find the optimal amount of detergent to use, and then automatic feed equipment can be installed. It is quite likely that the equipment can be paid for in a short time just by the cost of detergent saved. The use of less rinse water reduces waste volumes. There will be less tendency to emulsify tramp oil in the system, and the total dissolved solids will be lower. These are all real pluses in time, money, and trouble saved.

When one is faced with an unavoidable emulsion, time should be spent determining where the best place in the process is to break it. If the solids are dropped out first, will they entrain the oil? Will the free oil separate cleanly without interfacial sludge? Will adjustment of the pH give a cleaner initial separation? Does temperature have an effect? Is the feed constant, or are there frequent flow-surges? Are there random periods when large amounts of certain materials are dumped? If the volume of these deviant flows can be gauged, it may be possible to design the system so the first stilling pond will easily hold the largest of these events. Then you can bypass the first pond and treat the excess batchwise. Where justified, sensors can be installed in the input to detect and divert wild flows. Such procedures are well worth considering.

Whenever an "interface" tends to build up, it is usually better to try to treat before this step to prevent the interfacial accumulation than it is to isolate the interfacial sludge and try to treat it. If the sludge accumulation is slow and the rest of the system works well, it may be best to collect the sludge and treat it batchwise. In this case the help of a vendor of emulsion-breaking chemicals may be well advised. Field men for the makers of emulsion-breaking chemicals may have encountered similar problems before and have just the right answer. They can save a lot of lab time and headaches.

## CHEMICALS USED IN DEMULSIFICATION

Numerous chemicals used in demulsification are being marketed, and a few examples are listed below. Most of the emulsion-breaking chemicals used in the breaking of oil-in-water emulsions either adjust the pH, supply large cations, or are polymers which function as flocculants. Large amounts of lime, alum, ferric chloride, and clays are used to form flocs or adsorb small amounts of oil. Sand, gravel, hay, and other porous media are used in packed-bed coalescers. Clays are used as filter aids or as precoats as are diatomacious earth, wood flour, and sawdust.

Flocculant polymers are usually high-molecular-weight polymers derived from ethylene, propylene, vinyl, vinylidene, vinylacetate, and similar polymers. Acidic waters may be neutralized with limestone, lime, caustic soda, or ammonia. Basic streams are treated with waste acids such as plating wastes, spent sulfonation-acid, and sulfuric or hydrochloric acid recovered from scrubbers.

None of these materials is really a specialized demulsifying chemical. They serve to neutralize the charge on the oil drops or to help the oil drops to clump together. An agent to help destabilize the surface film or aid in coalescence may still be needed. A general, somewhat helpful rule of thumb is that those substances which stabilize an oil-in-water emulsion will tend to break a water-in-oil emulsion and vice versa. Also, materials which will react with the emulsifiers to change their solubility in either phase will tend to break the emulsion. Thus, an oil-in-water emulsion stabilized by sodium stearate can be broken by adding enough calcium ion to form calcium stearate, which will dissolve in the oil and thus destabilize the emulsion. If you then skim off the oil and some of the water at the interface and pump the sludge to another tank, you may find that the calcium stearate has emulsified the water into the oil. In the case of emulsions of valuable oils that are insoluble in water, it may be possible to add small amounts of a water-soluble alcohol or ketone so as to cause some of the emulsion stabilizer to go out of the interface into the water phase. Again, too much cosolvent may extract some of the material you want to recover into the water.

## SUMMARY

We said at the outset that demulsification is an art as well as a science. Indeed, when a demulsification formulation really works, it sometimes looks like black magic. But one man's magic is another man's technology. We are beginning to learn what the factors are that contribute to the formation and stabilization of emulsions. Unfortunately, we often cannot control the variables that we know are significant. We have to take them as they come. We know that we need to neutralize the charges on the drops, but we know we mustn't overtreat. We know that polymers and polyfunctional materials will help gather the drops into clumps. We know that we then have to destabilize the interface so that coalescence can occur. We have learned that this can be done by changes in temperature and by other physical effects. We know that in some cases we can destroy the emulsifier by oxidation or by other chemical reactions. We know that we can change the solubility of the interface components by certain techniques. We know that in some cases we can displace the stabilizers or dilute them with other surface-active materials. But we do not know in each and every specific case what the particular stabilizers or their concentrations are. Often they are complex mixtures. We often do not have time to do detailed analyses. We have to break the emulsion and proceed. This is where experience is essential. Professionals in the area of demulsification can draw on many years of experience in testing and breaking emulsions. They have a library of compounds and blends which have been found useful in similar cases. This is their stock-in-trade. Once you have as complete an idea as possible of the problem facing you, it often pays to call in an expert in the field. Let him practice his art and solve your problem.

## REFERENCES


1. I. L. Mongait, *Tr. Vses, Nauch-Issled.* S. B. Zakhar' ina, and E. B. Ishkhanova Inst. Vodornabzh., (1970) *28*:67-71
2. H. Espenhorst, Ger. Patent 1,442,386 (1959)
3. J. H. Dailey, Ger. Patent 2.042,310 (1971)
4. R. Chantereau, *Cent. Doc. Sidering Circ. Inform. Techn. Series 27*:2711-2722 (1970)
5. P. Mathews, Brit. Patent 1.192,541 (1965)
6. G. H. Wyatt, Trib. *CEBEDEAU 22*:419-423 (1969)
7. T. Oshima and S. Takayanagi, Jap. Patent 75-105,574 (1975)
8. J. Pilat, *Effluent and Water Treat., J. 14*:261-263 (1974)
9. V. F. Khroniykh, V. A. Livanov, and B. A. Mysjokin, *Medalwork Interfaces 2*:1219 (1977)
10. N. Spina, F. Naggi, and W. Arpesella, *Inquinamento, 20*: 39-44 (1978)
11. M. A. Ismail and P. G. Crandall, *Proc. Fla. State Hortic. Soc. 90*:156-158 (1978)
12. R. G. Luthy, R. E. Selleck, and T. R. Galloway, *Envirn. Sci. Tech. 11*:1211-1217 (1977)
13. E. G. Dressel, *Arch. Gewerbepath. Gewerbehyg. 11*:503-510 (1942)
14. T. R. Hazeltine, *Sewage Works J. 21*:859 (1949)
15. I. Pincus, K. F. Ockert, and C. R. Kinney, *Ind. Eng. Chem. 43*:521-524 (1951)
16. S. S. Voyulskij, N. M. Fodiman, and R. M. Panich, *Izvest. Vysshikh. Ucheb. Zavadinii, Khim. i Khim. Teknol. 2*:170-179 (1958)

17. H. Gummels, *Gesundh. Ingr.* *79*:1617 (1958)
18. J. B. Coulter, F. M. Crompton, and J. F. Langese, *Proc. Ind. Waste Conf.* *11*:99-113 (1957)
19. S. Castorina, *Olivicolt* *11*:4-7 (1956)
20. A. R. Balden, *Sewage and Industrial Wastes* *31*:934-940 (1959)
21. H. V. Farr, C. D. Harrington, J. A. Kyger, R. R. Midlam, W. C. Rosenbaum, H. E. Thayer, H. L. Wibbles, and H. Yeager, *U.S. At. Energy Comm.*, A-1084, 34pp. (1954) declass (1959)
22. P. L. Douglas, U.S. Patent 3,464,976 (1969)
23. *API Manual on Disposal of Refinery Wastes, Volume on Liquid Wastes*, API, New York
24. *Process Design Manual for Suspended Solids Removal*, EPA 625/1-75, U.S. EPA 1975
25. *Wastewater Treatment Plant Design, Water Pollution Control Federation*, Washington, D.C. 1977
26. P. L. Douglas, U.S. Patent 3,463,976 (1966)
27. T. Samocha, Polish Patent 94,318 (1977)
28. V. M. Skul'sku, S. V. Militsin, and P. Y. Kohodorovskii, *Lakokras. Mater. Ikh. Primen.* *2*:64-66 (1979)
29. H. Schilinski, R. Fichtmueller, and J. Tscherner, *Schmierstoffe Schmierungstech* *17*:34-38 (1967)
30. T. D. Henley, U.S. Patent 4,136,217 (1979)
31. V. D. Beaucaire, U.S. Patent 3.986,953 (1974)
32. J. E. Barker, V. W. Voltz, and R. S. Thompson, Chem. Eng. Progr., Symp. Ser. 1971, 67(107), 423-427, C.A. 74: 146070x
33. H. Weidmann, *Wasser Luft. Betr.* *14*:182-188 (1970)
34. T. G. Somer, *J. Appl. Chem. (London)* *13*:334-339 (1963)
35. Aktieselskapet Apothekernes Laboritorium for Special praeparater Neth. Pat. Appl. 6,614,914 April 24, 1967
36. G. Lauermann, J. Musner, and J. Schulz, Ger. (east) Patent 134,192 (1979)
37. P. Sole, Jr., U.S. Patent 3,451,990 (1979)
38. W. H. Hoffman, U.S. Patent 4,040,955 (1977)
39. S. Takeneshi, *Jpn. Kokai Tokkyo Koho* (Published, Unexamined patent application) 78-132, 782 (1978)
40. D. D. Davidson, U.S. Patent 3,350,301 (1967)
41. J. N. Chieu, E. F. Gloyna, and R. S. Schechter, *Proc. Ind. Waste Conf. 1975* *30*:611-622 (1977)
42. J. R. Stausser and R. S. Kurland, U.S. Patent 3,507,782 (1967)
43. H. Sedlmeier and H. W. Thoenes, *Tech. Mitt.* *64*:261-264 (1971)
44. C. D. Rurness, J. T. Allenson, Ger. Patent 1,803,229 (1969)

45. K. Nakai, Jpn. Tokkyo Koho. 79-0363 (1979)
46. M. L. Hemming, Brit. Patent Appl. 2,001,542 (1979)
47. L. D. Megal, U.S. Patent 3,432,426 (1969)
48. M. H. Weintraub, R. L. Gealer, and M. A. Dzieciuch, Ger. Patent 2,757,776 (1976)
49. R. Loison, *Eau Ind.* *20*:73-74 (1977)
50. Anon, *Ind. Miljoe* *20*:1,5,7,17,19 ff. (1974)
51. A. Gaser, Ger Patent 1,926,623 (1970)
52. W. Crist, J. Cyprian, and A. Gaser, *Wasser Luft. Betr.* *12*:755-757 (1968)
53. M. Z. Mavlyutova, and I. Asafagan, *Nefteprom. Delo 4*: 3509 (1969)
54. K. J. Lissant, *J. Colloid Interface Sci.* *22*:462 (1966)
55. K. J. lissant, *J. Soc. Cosmet, Chem.* *21*:141 (1970)
56. K. J. Lissant and K. G. Mayhan, *J. Colloid Interface Sci.* *42*:201 (1973)
57. K. J. Lissant, K. G. Mayhan, and B. W. Peace, *J. Colloid Interface Sci.* *47*:416 (1974)
58. K. J. Lissant in *Emulsions and Emulsion Technology*, K. J. Lissant, (ed.) *Part 1*, M. Dekker, New York, 1974, pp. 102ff
59. H. M. Princen, *J. Colloid Interface Sci.* *71*:55-66 (1979)
60. J. M. Princen, M. P. Aronson, and J. C. Maser, *J. Colloid Interface Sci.* *75:*:246-270 (1980)

# 4

# Demulsification of Water-in-Oil Emulsions

## INTRODUCTION

This chapter will cover the occurrence, avoidance, and treatment of water-in-oil (W/O) emulsions in general. Detailed discussion of emulsions in the petroleum industry will be discussed in Chapter 5. Water-in-oil emulsions are not as common as oil-in-water (O/W) emulsions. They do, however, occur much more frequently than is usually thought. Often they are not recognized as emulsions. "Cloudy oils", "wet oils", "sludges", "interfaces", "skimmings" are often names under which water-in-oil emulsions are often hidden. In the production of vegetable oils, the macerated plant material is pressed to obtain the oil. If the pressing are cloudy, they may be treated with a small amount of clay and then filtered. This process often produces a "bright oil". What we have done in many cases is to break a water-in-oil emulsion by adsorbing the water on the clay particles. Sometimes the fact that we have broken an emulsion is further hidden by the fact that the clay was not added to the oil directly but was used as a precoat on a filter where it performed the same function. In this case, however, we can encounter difficulties. If too much oil is put through the filter, the capacity of the clay precoat to adsorb water may be exceeded. Suddenly, the filter seems to be passing solids when, really, it is letting fine water-particles through. If there is too much water, the clay may swell. The filter then "blinds" rapidly.

The use of adsorbants to remove water from oil is an old and time-honored process. In cases where the amount of water is small, it is often the easiest way. But disposal of the oily residue or filter cake must be planned. Several practical ways to use the material have been found. In the case of vegetable oils the press cake is often used to make excellent cattle food. Reasonable amounts of the filter cake can also be incorporated into the press cake. The oil adds nutritive value to the animal food. In the case of inedible oils, the clay can sometimes be incorporated into such things as "charcoal" briquettes and burned. In some instances the filter cake can be purged with steam, dried, and reused. The oil can be recovered from the steam condensate.

It is easy to see that if anything is added to an emulsion to break it, one has to consider where the material will go in the separation process. This problem was brought out by the use of inorganic flocculants in the treatment of waste water. The use of flocculation or flotation processes is widespread in the water-treatment field. The use of lime, alum, and iron compounds for this purpose runs to millions of tons per year, all of which must be disposed of. In the cases where significant amounts of oil are recovered with the froth or skimmings, a real problem can develop. We have seen how a concentrated oil-in-water emulsion can be inverted into a water-in-oil emulsion by pumping or other transfer processes. If we now try to break this invert emulsion, we have to contend with the presence of large amounts of solids, usually at the interface. More and more operators of treatment plants are, therefore, going to the use of polymeric flocculants which will allow a drastic reduction in the amounts of inorganics required. The polymers themselves are used at very low concentrations and can be degraded by shear. The avoidance of large volumes of inorganic floc greatly simplifies the recovery of oil from the skimmings, sludge, or froth.

## BREAKING METHODS

Now, how does one break a water-in-oil emulsion? In these emulsions the effects of pH or salts are much less than in their oil-in-water counterparts. In some cases the addition of oil-soluble acids such as acetic acid has been used to destabilize emulsions. Washing of the oil with additional fresh water has also been used to scavenge small water drops in oil. The fundamental problem in treating water-in-oil emulsions is getting the treatment chemicals to the oil-water interface. In general, the oils are more viscous than water, and the transport of materials through the oil is slower. The oils are, as a class, un-ionized and nonconductive. Inorganics are not soluble in most oils, and, thus, are difficult to use to treat water-in-oil- emulsions. On the other hand, since the oils are nonconductive, electrical fields can be used to cause coalescence of the drops of water. Cottrell patented this concept long ago [1]. The use of heat to break water-in-oil emulsions is widespread. Heat treatment is practical since the oil has a much lower heat capacity than water. Less energy is therefore needed to gain a particular temperature. In his book on the technology of fats and oils Kirschenbauer states, "Leaving the mixer, the charge is passed to a heating unit. Here the temperature is rapidly raised to 55-70°C to break the emulsion which invariably forms in the mixer" [2]. As a matter of fact, in animal and vegetable oil production the wet oil is usually steam stripped or vacuum stripped to remove small amounts of water that are dissolved or emulsified and would separate on cooling. The use of heat to break water-in-oil emulsions is also effective because heating reduces the viscosity of the oil and also reduces the density of the oil faster than the density of the water phase. (Oils have larger coefficients of thermal expansion than water.) The elevated temperature also increases the solubility of the surface-active materials in both the oil and the water and, thus, tends to weaken the interfacial film. All these effects combine to make heating one of the preferred methods of breaking W/O emulsions.

Two other phenomena are useful in breaking these emulsions. The first is the addition of polyfunctional oil-soluble materials that will serve the same function as the polymeric polyelectrolytes in the aqueous systems. These compounds will be adsorbed at the interfaces of the drops and tend to gather them into clumps where coalescence can occur. The clumps will also settle faster. The second phenomenon is that of disrupting the organization of the oil-water interface. In some oil systems the surface of the water drops tends to become almost skin. Any hydrophilic material in the oil is scavenged by the relatively small amount of water, and the interfaces become cluttered with partially soluble or insoluble species. The presence of this trash tends to keep the interfaces of adjacent drops out of contact and slows or prevents coalescence. In water-external emulsions it is often possible to alleviate such conditions by adding something that will promote waterwetting of the interfacial material and pull it into the external phase. The promotion of oil-wetting in the oil-external emulsions is harder to accomplish and often undesirable. Forcing the material into the small water drops may not help. Sometimes the addition of more water with a wetting agent, followed by controlled agitation, will bring the debris into the water phase as settling occurs. The addition of a cosolvent for the interfacial material may drive it from the interface or help it to dissolve in the oil and break the emulsion.

It is often not clear what the stabilizing agents for a water-in-oil emulsion are. Almost all naturally occurring oily materials are complex mixtures. They are often subject to oxidation and the attack of acids or bases. The wide spectrum of decomposition and reaction products tends to consist of surface-active materials. They may be present in concentrations ranging from parts per million to several percent. This is one of the reasons why the breaking of water-in-oil emulsions is called an art. Over the years a backlog of experience has developed in each industry. It is often of a very empirical nature. A number of examples may be of use. A large steel mill had an accumulation of oily wastes consisting of a water-in-oil emulsion containing from 50 to 80% water. This material was successfully treated by a batch process whereby the emulsion was heated to 160°F, and 5000 ppm of a commercial demulsifier were added. The tank was rolled with air for 2 hr and allowed to stand for 24-48 hr. The water content was reduced to about 1%, and the oil was then equal in quality to No. 6 fuel oil. The oil was used as fuel. It was estimated that the treatment costs were about one-quarter of the cost of the fuel recovered. The recovered water was able to meet specifications for discharge with about 15 ppm solids and only a trace of oil.

Another steel mill had a waste oil consisting of hydraulic oils, cutting oils, lubricating oils, transmission oils, and some untreated rolling-oil emulsion. This waste oil was run through an American Petroleum Institute (API) separator, and the skimmings went to a hot-oil sump and were then pumped to an oil storage tank. Bottle tests were performed on 100-ml samples of the slop oil using several commercial demulsifying formulations. One was found that results in clean top oil, very little interface, and substantial water drawoff.

An automotive plant had a sludge treatment and disposal problem. The addition of a polymeric flocculant eliminated the need to add alum, thus reducing the volume of the sludge by two to three truckloads per week. It was then possible to recover oil from the sludge and, by treating with a demulsifier, to recover 7,000-10,000 gal of oil per week. The changes in operations were estimated to be saving about $20,000 per month. Another plant expects to save $200,000 per year by adopting a similar procedure.

A British company was experiencing trouble recovering solids and oil from a waste stream. On-site testing showed that the use of a commercial tank-bottom demulsifer in conjunction with pH control resulted in the recovery of clean oil, fit for fuel. It was also found that enough mill scale could be recovered to allow it to be sold back to a steel mill. The capacity of the system was increased to the point that waste oils from other sources were being considered for treatment. It was estimated that about 12,000 gal of usable fuel oil were being recovered per week.

In another case it was found that there were two major costs in the disposal process: those of treatment and of hauling the sludge. Although the use of organics instead of lime and alum increased the treatment costs, the saving in the reduced amount of sludge that had to be hauled more than offset the cost of the chemical. A further advantage was the fact that the capacity of the system was increased to the point that the building of larger facilities could be avoided.

A common thread runs through all these examples. The situation first involved an oil-in-water emulsion; then the oily skimmings or sludge had to be treated as a water-in-oil emulsion. In many cases the presence of large amounts of inorganic solids obscured the fact that we were fundamentally dealing with a water-in-oil emulsion.

It should be becoming clear that we cannot really separate one type of emulsion from another in field practice. The possibility of both should be considered in any planning. In this chapter we have arbitrarily eliminated from consideration the emulsion problems of the petroleum industry. There is good reason for this, but it is well to remember that the other industries have learned from the petroleum industry. The use of the API separator is a good (and sometimes bad) example. We will, therefore, spend some time discussing possible mechanisms for the breaking of oil-external emulsions at this point and then assume that they will not need to be discussed in the following chapters.

Most of the work on emulsion stability has been done with the purpose of producing stable emulsions. Nearly all of the literature uses a few classical oil-in-water prototype emulsions and tries to stabilize them. This approach is not much help to us in demulsification. In the first place, most of the classical emulsions use true soaps as the emulsifiers. Many of the present surfactants were not known when the first work was done. After a study of the fairly simple soap emulsions had been completed, it was assumed that the work had all been done. The field was almost abandoned for many years. The more recent work has also been directed towards the production of stable emulsions--in many cases, microemulsions. Details of the use of specific materials for breaking emulsions are confined for the most part of the Russian literature and to the patent literature. Most of the literature is specifically directed to use in breaking petroleum emulsions. Almost none of the published material directs itself to the question of the mechanisms of the water-in-oil demulsification process. Lawrence [3] writes of some problems encountered in breaking water-in-oil emulsions in the fuel tanks of naval ships. He has some rather unkind words for some of the previous authors. He says,

> "There was obviously a very efficient emulsifying agent, but no clue as to its nature. We therefore set out to determine quantitatively what role an emulsifier has to play. The text-books have no hesitation: they tell us that the interfacial tension must be lowered and that a high interfacial tension is a tendency to coagulation. This argument is . . . stupid. . .

He then goes on to express an interesting idea. He says that coagulation involves two processes: contact and coalescence. Interfacial tension has nothing to do with the collision frequency of the drops. It is involved in coalescence. He says,

> The first step in coalescence must involve distortion and an increase of interfacial area. The lower the value of interfacial tension, the less work will be required for this step in coalescence. In other words, *low* interfacial tension would be expected to promote coalescence. That is, the condition laid down by the textbooks for stability.

Lawrence then goes on to analyze the effect of Brownian motion on the collision frequency [3]. He does the same for the Stokes' settling. We have covered these items in Chapter 1. Having taken care of the collision frequency, Lawrence then goes on to discuss coalescence. He raises the question of the function of the emulsifying agent. He points out that except for its possible effect on the viscosity of the external phase, it can have no effect on the collision frequency. Its action is to prevent coalescence.

> It forms an energy barrier at the interface. This may be electrical where the like charges on colliding droplets cause mutual repulsion: it may be mechanical in those cases where the emulsifier forms an elastic membrane around the droplets; or it may be one due to surface energy in those emulsions stabilized by a powder differentially wetted by the two liquids.

He then discusses the effect of various variables in overcoming the energy barrier. He shows how an increase in temperature increases Brownian movement and decreases the viscosity, how centrifuging can change the G factor in the sedimentation equation, and how all of these things do not change the value of the energy barrier. He then points out that our best effort in breaking an emulsion is to reduce the energy barrier. If this barrier is mechanical, we have to attack the "membrane". This can be done by heating or by "adding a 'fluxing' agent, such as cresol, which will reduce the rigidity of the 'elastic bags' around the water drops." He next points out that good fluxing agents tend to be soluble in the oil and must, therefore, be used in large amounts to be effective. He suggests that the addition of a water-soluble surfactant that is more surface-active than the emulsifier will tend to displace the emulsifier from the interface and destabilize the emulsion. Since the water is present in lower amounts, less agent will be required. He cautions against using an excess of agent because it might cause inversion. He also suggests that increasing the temperature at this point will increase the solubility of the emulsifier in the oil and aid in driving it from the interface.

The use of other additives to increase the solubility of the emulsifier in the oil is also suggested [3]. One more portion at least is worth quoting.

> Lowering of interfacial tension is not the factor which makes a substance an emulsifier; surface activity is necessary for the substance to form the energy barrier, but the barrier is electrical and mechanical. It is the physical properties of the adsorbed layer which decide whether the adsorbed film will be an emulsifying agent or a demulsifier. . . . This difference of the properties of the adsorbed film is due, not only to the nature of the molecule, but also to its concentration and to its temperature, hence we have the position that a substance may be good demulsifying agent at low concentration but a good emulsifier at higher concentration.

All this material was written in 1948, yet it is as true today as it was then--and as incomplete.

## CHEMICAL DEMULSIFICATION

We still don't know in many cases what the exact emulsifiers are in a specific system. But we can still approach the problem from the same basis. We must remember that an emulsion is a construct and that it has certain properties because of its structure and physical conformation. It is this structure that sometimes must be physically broken before the emulsion can be resolved. It is well to remember that most demulsifiers in water-in-oil emulsions are not very soluble in either the oil or the water phase. We are thus faced with the problem of delivering them to the interface. To do this it is sometimes useful to dissolve them in a solvent which is miscible with oil. It is then necessary to add the demulsifier solution and mix it thoroughly with the emulsion before the oil phase has a chance to dilute the solvent and cause the demulsifier to come out of solution. Thus, in the use of any demulsifier it is necessary to be sure that the treatment chemical is added in such a way that local excesses are avoided and so that the demulsifier can contact the individual drops. It is also necessary to allow time for the demulsifying agent to enter the interface and displace or modify the surface film. Additional time is required for the drops to meet, coalesce, and settle. This is why it is common practice to meter in a chemical just ahead of a pump, and then let the emulsion travel down a line before reaching a settling tank or separator. If the treating agent is not mixed in well, there may be local overtreatment and incomplete separation, or more chemical may be required than is really necessary. If enough time is not allowed for "ageing" before it reaches the separator, some of the material may pass through the separator before resolution is complete.

Bottle tests and pilot-plant tests will show the amount of mixing required to get results and the rates of separation and settling that can be expected. The effect of concentration and temperature can also be assessed. The danger of overtreatment can be evaluated, and the presence of interface detected. It is important to remember that how the chemical is added is as significant as what is added.

In batch treatment of emulsions it is particularly important to mix the chemical in thoroughly. Dumping in the whole amount and then "rolling the tank" can often be made to work, but it may be possible to do the job better if the agent can be mixed in well as it is added. Not only is it necessary to get an even distribution of the chemical in the tank, but it is also necessary to avoid making the emulsion worse by stirring or mixing too hard or too much. Just as in flocculation of solids in oil-in-water systems, it is necessary to have a mixing stage, then a gathering stage, and then a settling stage. The destabilizing agent should be thoroughly dispersed so that it can contact all droplets evenly. The droplets should then be kept in suspension for a short time to allow them to come into contact and the droplet size to grow. In this stage also, the larger drops can scavenge the smaller drops and eliminate them.

Now there should be a period of relative quiescence so that the drops can settle and form a separate phase. In the case of water-in-oil emulsions the water drops will usually settle to the bottom of the vessel and form a water layer. If the correct procedures have been worked out, the large drops will have swept up the small ones, and a clean break with dry, bright oil on top and a sharp interface will result. Even if the right demulsifier has been used at the optimum concentration, it is still possible to get poor results if the mixing is improper or if most of the water is allowed to fall out before the very small drops can be swept with it.

As the drops coalesce with each other and with the water interface, the total interfacial area decreases, and any insoluble material that is left at the interface is crowded together until it may constitute a mechanical barrier to further coalescence. When this happens, a complex interfacial layer develops. These interfaces can be very difficult to resolve. Even if the concentration of insoluble material in the original emulsion is very low, substantial volumes can accumulate in normal industrial treating plants. It is necessary, therefore, to try to provide a treating agent that will remove these materials from the interface. This can be done by selectively wetting them and driving them into the oil or the water phase. If the materials are combustible and the oil is to be used as fuel, it may be acceptable to oil wet the interfacial materials. If not, they may have to be dissolved in or dispersed into the water.

A successful emulsion-breaking formulation must be multifunctional. It must be a single phase that is stable over the ranges of temperatures encountered in shipping, storage, and use. It must be fluid enough under these conditions to be fed to the emulsion stream accurately. Its composition must retain the active materials in solution in the oil long enough to let them migrate to the interface. It must contain materials to neutralize charges, materials to destabilize adsorbed films, materials to aid in gathering the drops, and materials to selectively wet any solids. It must meet these constraints without causing a tendency for the emulsion to invert. No wonder some of the formulas appear to be quite complex. Indeed, almost any type of chemical compound that can be named has been suggested as a demulsifier. What's more, at least one instance of effectiveness can usually be found for any type of material. But the selection of the best material and the blending of materials of differing functions to make a balanced formulation is still very much an art. It is often possible after the fact to explain how a particular material worked in a particular case, but the predictive capacity of these explanations to new cases is poor. In spite of this, it is a fact that individual treating engineers do seem to develop a feel for what should be used in certain instances. They seem to be able to put together facts from their experience and to arrive at a selection for a given case faster than someone without the experience. There is, therefore, reason to believe that behind the mountain of field data there are some unifying principles. It is these principles that we want to try to identify and use. It would be nice if we could discover and develop some sort of "demulsification index", an equation that we could fit values into and solve to come up with the right formulation for each new application.

We have already, in this book, begun to outline some of the factors that must be considered. What type of emulsion are we dealing with? What is the phase ratio? What is the particle size distribution? What is the gross composition of each phase? Are there any solids present? Are the solids water wetted or oil wetted? Do we know specifically of any surface-active materials in either phase? Are these materials water-soluble or oil-soluble? Are they cationic, anionic, or nonionic? How does their solubility in each phase vary with temperature, pressure, ionic strength, pH? Is the composition of the stream to be treated constant, or does it vary? How much does it vary, and how fast? At what rate must we be able to treat the stream? How complete must the separations be. Are we required to keep the material chemically pure? Are we dealing with a food product? A drug? Is the stream toxic or corrosive? What are the economics of the process? How do we utilize or dispose of the residues, sludges, etc? What mechanical procedures have been performed on the stream? Has it been pumped, filtered, exposed to oxygen, subjected to high shear? Has the temperature been cycled? Has the pressure changed? What equipment is available for treating this stream. Is this a long-term problem or a one-time condition? Can the situation be avoided in the future?

It is clear that there are too many variables, many of them uncontrollable, to allow a neat, mathematical solution to the general case. Enough experience has been accumulated in the more common cases, however, to allow us to have come up with reasonable answers in certain fields. The experience from these fields can be applied to other areas if care is used. There is also a need to take a fresh look at all of the applications to see if new technologies can be of help. The availability of new polymers and other organic moieties opens new fields. Ion exchange and membrane techniques are coming of age. An item in the May 1981 edition of "Chemical Processing" reports that an ultrafiltration unit is being used to extract 90,000 gal/day of pure water from a dilute oil-in-water emulsion [4]. The reclaimed water is recycled to the process and

". . . the concentrated emulsions can be further processed." We have seen that it may be easier to handle a small volume of concentrated emulsion than a large volume of a very dilute one. It is well to note that the phase ratio of the "concentrated emulsion" is not given. It may not be what we would call concentrated.

One of the ways of treating water-in-oil emulsions that we have just mentioned is electrical treatment. Since the external, oil phase is not conductive, a voltage difference can be imposed across electrodes without drawing large amounts of current. The field so imposed will cause the water droplets to migrate to the electrode of the opposite charge and coalesce. Even if the droplets were not originally charged, the electrical field will induce a charge and attract and distort them. If electrical treatment can be used alone it has the advantage of not adding anything to the stream. The process can be used in systems where contamination must be avoided. Electrical treatment will be discussed in detail in Chapter 5.

## SUMMARY

In summary, then, we can say that compared to oil-in-water emulsions, water-in-oil emulsions are:

1. Less sensitive to pH
2. As sensitive or more sensitive to heat
3. Amenable to electrical treatment
4. Except for the petroleum industry, relatively rare
5. Often present but unrecognized in sludges, interfaces, scums, and froths

## REFERENCES


1. F. G. Cottrell, U.S. Patent 98,114 (1911)
2. H. A. Kirschenbauer, *Fats and Oils, Med.* Reinhold, New York p. 73
3. A. S. C. Lawrence, *Chemistry and Industry 39*:615-617 (1948)
4. Anon, *Chem. Process* (Chicago) V. 44, p. 186

# 5

# Demulsification of Petroleum Emulsions

## INTRODUCTION

In terms of time, money, and effort the only real place where emulsions are recognized as a problem is in the petroleum industry. In other industries there are emulsion problems, but in the petroleum industry, demulsification is a major subindustry. Hundreds of millions of dollars are spent each year in preventing or breaking petroleum emulsions. This industry gave us a plethora of excellent examples of how to deal with emulsions and how not to deal with them. As mentioned before, the rest of industry has learned from the petroleum industry in breaking emulsions.

## ORIGIN AND OCCURRENCE

How and where are petroleum emulsions formed? Some are present in the oil reservoirs; some are formed during the recovery process. There are those who say that no emulsion exists in the undisturbed oil reservoir but that emulsification begins after production starts. It is doubtful that anyone has ever been able to examine a really undisturbed crude-oil reservoir, but studies have been made on cores taken from production zones. The results are inconclusive. Some cores have been found with emulsion in them; some do not have emulsion. It is probably another case of the wide variability among oil deposits.

It may help to review for a short time the situation in an oil deposit and the methods used to get the oil out of the ground. Figure 5.1 is a drawing of a typical oil field. Even the experts do

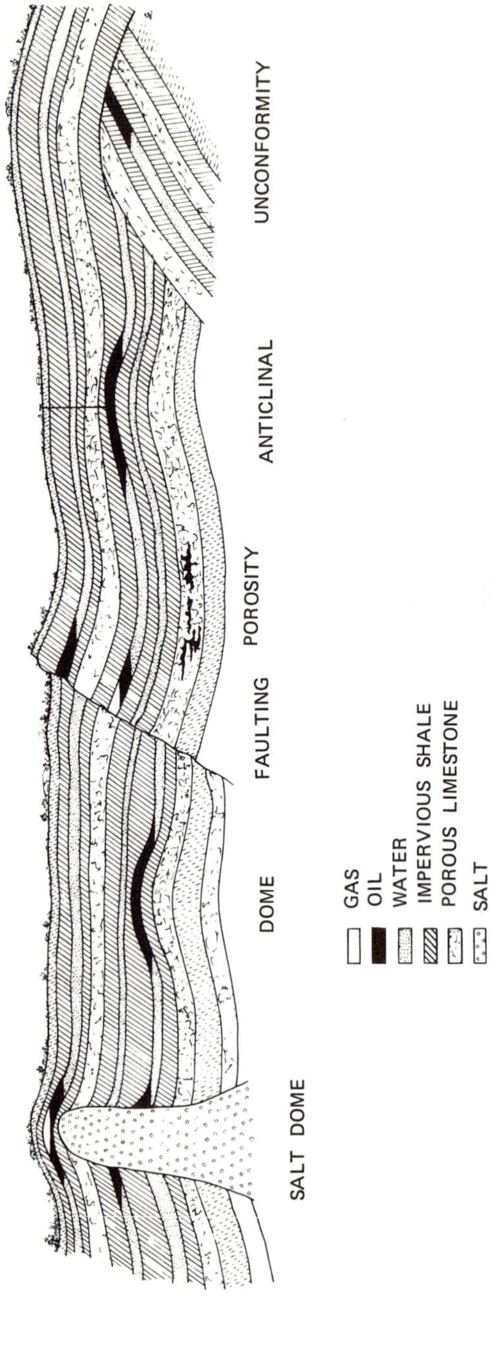


Figure 5.1 Types of natural oil reservoirs.

not completely agree on the mechanisms of the formation of petroleum or the origin of oil deposits. It is gererally held that deposits of organic material, probably of marine or at least aquatic origin, were laid down and covered by layers of mud, silt, and sand. These materials were then subjected to heat and pressure as the layers of rock above them formed. Other experts feel that the action of heat and pressure on inorganic materials deep in the earth may have produced gases and hydrocarbons. In any event, whatever the original production site or mechanism, the material was probably produced in small amounts over a fairly large volume of strata. The oils and gases then migrated, impelled by density differences, pressure and temperature gradients, and, perhaps, actual tectonic movements, into and through various porous formations. If conditions were right, the materials could then accumulate at points where their movement was impeded and "pools" of petroleum could form. Several possible "traps" are depicted in Figure 5.1. It must be made clear that the oil does not reside in holes or caves. It is distributed in the fine pores of the rock or in the interstices between grains of sand. Typical cores from producing oil-wells will seldom show large holes. "Vuggy" formations, like Swiss cheese, are not really common. At the same time, there is also brine or water in the pores. Even in the formations that are almost completely filled with oil, the actual rock surfaces may still be substantially water wet.

When an exploratory well is drilled, it may penetrate one of these pockets of oil-filled rock, and oil can then flow into the well bore. The flow of oil into the well can be induced by several factors. The overlying strata will press down on the oil formation, and this pressure may be enough to move the oil. If there is a pocket of gas under pressure above the oil, the gas will expand and cause the oil to flow. In some formations there is water below the oil, and this water can be compressed by the overburden. As the water expands, it will drive the oil. If the pressure is enough to overcome the static head in the well, we have a flowing well. In some cases as the pressure on the oil is relieved, gases dissolved in the oil will come out of solution, and the well will foam over like a hot bottle of soda. In other cases the oil has to be pumped out of the well. As petroleum is withdrawn from the formation, the pressure decreases, and, finally, the well has to be put on the pump.

There is some debate as to whether there is actually any emulsion in the formation or whether the emulsion forms as the oil is produced. Conditions vary so widely from field to field and from well to well that it is most likely that both situations can be found.

One of the most common oil-field situations is one where the pool of oil is underlain by a pool of water or brine. When wells are drilled into the formation, some will penetrate only the oil region; some will hit only the water region (These are still called "dry holes"); and some will first hit the oil and then, as they are drilled deeper to determine the depth of the "pay", the water section. As the crude oil is produced, the water level will rise, and some wells will be flooded. If the oil is withdrawn from the formation too rapidly, the water may "cone", and the well starts to make wet oil. Besides entering from the oil-bearing formation, water can also get into the well from strata that were penetrated before hitting the pay. If these zones are not carefully cased and sealed off, they can present difficulties.

If water is being produced with the oil and all practical methods have been employed to exclude the water, we can still minimize the emulsion problems by avoiding situations where the oil and water will be subjected to high shear. Carefully placed chokes to prevent rapid pressure drops can be of great help. Partially closed valves should be avoided as should worn pump-parts, leaky packers, or slapping sucker-rods.

## BOTTOM SETTLINGS AND WATER TESTING

Even if all these precautions are taken, there still will be emulsion problems. What approach should be followed to resolve the emulsion? Most of the emulsions that occur in oil fields are of the water-in-oil (W/O) type. There may be from 1% to as much as 60 or 70% water dispersed in the oil. Some of this water will settle out rapidly and is known as "free water". The rest is classed as bottom settlings or basic sediment (BS). Before crude oil is accepted for transmission down a pipeline, it is tested for bottom settlings and water (BS & W). This test is done by placing 50 ml of the oil and 50 ml of a solvent into a conical-bottomed 100 ml centrifuge tube, which is graduated to read directly in percent BS & W. The liquids are mixed and then centrifuged for 2-10 min. The tubes may be heated in some cases. The interface between clean oil and the other materials is read to determine BS & W. Pipeline limits for acceptance of crude range from 0.2 to 3% BS & W depending on the field.

If the oil exceeds these limits, it must be treated before it can be shipped. In the best of cases, simple, room-temperature, gravity-settling for a few hours will release enough of the free water to meet pipeline specs. If not, other means must be employed.

These methods can be broken into three general classes: Mechanical, electrical, and chemical. The choice of method depends primarily on economics, but all factors must be considered. Cost of materials, labor, capital investment, rate of throughput, and other factors must be taken into account when selecting a treating procedure. Often a combination of mechanical and chemical, or electrical and chemical is the best choice. The various methods will be discussed in some detail at this point. Remember that the basic principles that have been used all along still apply and that the specific examples discussed hereafter can often be applied to situations outside the petroleum industry. This is what is meant by "going to school" on the petroleum industry.

At this point we probably should introduce some more common oil-field terminology. These terms will help in searching the literature or in understanding articles in petroleum-industry periodicals. If the produced fluids from a well contain water or brine the oil is said to be cut. The water cut can be as high as 80%. If the water settles out easily it is free water. If the emulsion is easily broken it is a "loose" emulsion. A very stable emulsion is said to be "tight". If the emulsion is oil-in-water, it is called a "reverse" emulsion. If chemicals are injected into the well bore, they are said to be added downhole. This can be through the production string or down the casing.

## GAS SEPARATORS AND WATER KNOCKOUTS

Fluids produced at the wellhead consist of three phases: oil, water, and gas. It has been found that the presence of a gas phase in the system tends to increase the chances for emulsification. It is thought that this is due to the increased turbulence and shear induced by the gas phase. In any event it is advantageous to minimize the gas phase. It is also clear that the more quickly water can be removed from the system, the better. If nothing else, we save the cost of transporting it any further. These are some reasons why there are usually gas separators and "free water knockouts" near the wellhead. There are many different types and styles of separators. The conformations vary from field to field. Styles are dictated by the pressure of the produced fluids, the oil-water-gas ratios, the viscosity of the oil, and the presence of solids. Free water is defined as that water which will drop out at ambient conditions when the oil is held quiescent for 5 min. For this reason the knockouts are designed to slow the linear flow of the oil to a point where large water drops will fall out and to hold it there for about 5 min. They operate on a continuous basis with intermittent water blowdown. The gas separators are designed on a principle similar to the cyclone separators mentioned in Chapter 3. The fluid is introduced radially so that the liquid is thrown against the sides of the vessel and drains to the bottom. The gases then pass through a second section where they are "demisted" and are then sent to a gathering system. There may be provision for a water drawoff at the bottom of the unit. By its nature there is little turbulence in the water knockout, but there is significant agitation in the gas separators. If treating chemicals are to be added to the oil stream, they are often added before the water knockout to help its action and before the gas separator to insure good mixing of the chemical with the oil stream.

From the gas separators the produced oil then flows to units that will break any emulsion and remove the water. The simplest of these is a settling pond. This was much used in the early days of oil production but is little used today. If the oil would not separate in a settling pond or holding tank, it was run to holding ponds where the "roily oil" was allowed to stand and, periodically, any free oil was skimmed. This was known as "sunning". It was a wasteful and messy procedure. Most of the light ends of the oil were lost, and oxidation could occur. Today most oil production is kept in closed systems as much as possible.

## GUN BARRELS

Sometimes a water wash will help dehydrate an oil. Washing may be done in a "gun barrel" (Fig. 5.2). The gun barrel was so named because it consists of a long, slender, cylindrical tank standing on end. A pipe about one-quarter of the diameter of the cylinder enters the tank at the top and extends down the center of the cylinder to within about 2 ft of the bottom. The top of this pipe is closed and fitted with a gas drawoff. The bottom is open and attached to a "spreader". The spreader is usually a disk-shaped, inverted tray with perforations or notches in the periphery of the tray. The oil is introduced into the central downcomer just above where the pipe enters the main tank. Gas separates from the oil and collects in the top of the central pipe. The emulsion travels down the pipe and is distributed into a water layer in the bottom of the gun barrel. The drops of emulsion rise through the water layer giving the drops of water in the emulsion time to coalesce with the bulk water layer. The oil forms a layer at the top of the gun barrel from the top of which clean oil can be drawn (if

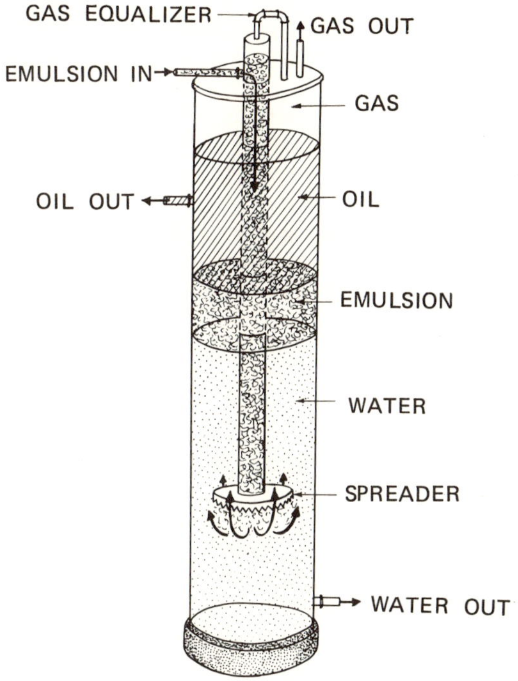


Figure 5.2 Typical gun barrel.

the process is working). Some gas may collect at the top of the gun barrel. It can be combined with the other gas and drawn off. Water is drawn from the bottom of the equipment. The water level in the gun barrel can be varied to suit the field conditions. If a longer wash period is needed, the level is held high. If more coalescence time is required, then the water level may be dropped.

## HEATER TREATERS

If simple holding tanks would not yield marketable oil, the next step was to try heating the oil. One way to do this was to pass the oil through heat exchangers before running it to the holding tanks. The increased temperature reduced the viscosity of the oil and often changed the solubility relationships of the natural emulsifiers in the oil so that the emulsion was broken and the water could be removed. The dehydrated oil was then either run to the pipeline or to storage tanks. The hot side of the heat exchangers was filled with steam, hot water, or the hot oil exiting the settler. Heat was provided by burning lease gas or oil. The next evolutionary step was the heater treater (Fig. 5.3). Here again there are many sizes and styles. They all have certain features in common. Details of the construction can be obtained from the manufacturers or from various American Petroleum Institute (API) publications [1-3].

A number of people over the years have written articles or booklets on oil-field demulsification [4-10]. Most of these articles seem to assume that the resolution method of choice is chemical treatment, usually in conjunction with heat or electricity. This is probably because no emulsion problem is thought to exist when purely mechanical methods are adequate for demulsification. If the

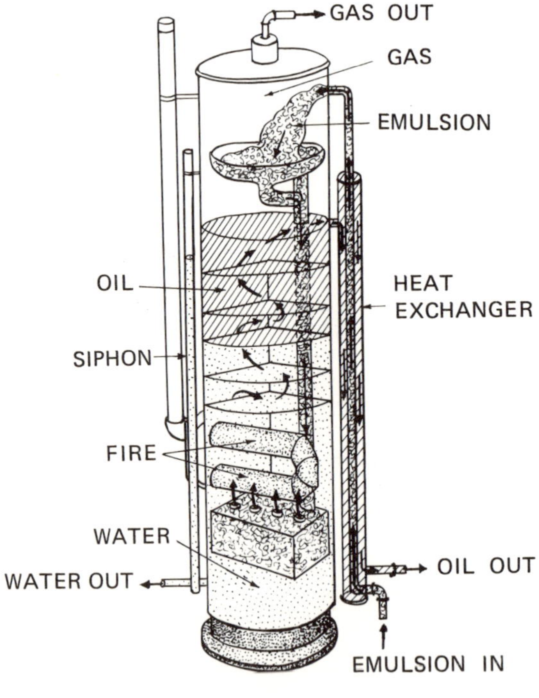


Figure 5.3 Heater treater.

water will separate well enough to meet pipeline specifications in a holding tank or gun barrel it is all free water, and there is no emulsion problem. If the nature of the production changes or if the volume of the fluid sent to the system increases, problems may develop. It is at this point that chemical treatment is tried in an attempt to make existing equipment do the job. Often this treatment is successful. This is probably the way that the heater-treater evolved. The vertical heater treater is in many ways a modified gun barrel. If a means is provided for heating the water in the gun barrel, or for preheating the incoming emulsion stream, we have a basic heater treater. If a chemical is used, it is added ahead of the treater.

The horizontal heater treater may have evolved from the horizontal pipe separators of the early years. It may also have resulted from an attempt to increase the interfacial area without increasing the volume of the vessel. Horizontal treaters are useful if wash times can be short and a large coalescence area is important.

## CENTRIFUGES

Centrifuges were used a good deal in the early days of the oil industry particularly in Kansas, Oklahoma, and some parts of the South. They are less used today. If the requirement for clean oil is a longer settling time, they may be useful. They essentially increase the value of G and the sedimentation rate. They do nothing for coalescence. Often centrifuges were used to get clean oil quickly. The BS & W which were produced from the centrifuge were treated by heat or chemicals. The use of centrifuges in conjunction with electrical treatment has also been noted [11]. Centrifuges encounter difficulty with very waxy crudes or on crudes where there are entrained solids. They have the advantage of being portable, but maintenance has been a problem at times.

## FILTERS

Filters have also been used to separate emulsions. They seem to operate in two ways. First, they crowd the water drops together through a small orifice and may thus aid coalescence. Secondly, they may hold back the water droplets until they can coalesce and fall out. Filter media can be either water wet or oil wet. If the filter medium is water wet, the medium must be quite coarse to allow the oil to flow through. The water is collected by the water-wet surfaces and can be made to drain out. Care must be exercised to prevent water being reentrained in the oil stream. Media packed coalescers such as "hay tanks" fall into this class. Oil-wet media will allow the oil to flow through and hold back the water droplets. They can then coalesce and fall out. Again, the flow rate must be controlled to prevent breakthrough. If solids are present, the filter may have to be backwashed or cleaned frequently.

## OTHER MECHANICAL METHODS

A number of mechanical methods of resolving emulsions have been patented [12-14]. These methods seem to stem from observations made in attempting to break emulsions on a laboratory scale. If some oil-field emulsions are heated or stirred and then allowed to stand in a beaker, a partial separation results. There will be a layer of relatively clean oil on top, a layer of water on the bottom, and, in between, a layer of loose emulsion. This emulsion has been called skins, lace, or other terms that convey, in part, its appearance. If this layer is stirred carefully with a glass rod or a stick, it will collapse and release most of the water. If the interface layer is rubbed between the hands, it can sometimes be broken. Many attempts have been made to imitate this stirring or rubbing action in a full-scale breaking system. Few of these attempts have been really successful. At least, few saw full-scale commercial acceptance. Fisher's patent [12] uses two gears or rollers--one water wet and the other oil wet. He suggests the use of this apparatus in conjunction with electrical treatment. Meinzer's patents [13,14] introduce finely divided solids--some oil wet, some water wet. This is, in a way, the same as a packed section except that the particles are loose in the liquid. Means are provided to separate the solids or to confine them to a specific portion of the apparatus. This should not be confused with the introduction of solids that can react with or adsorb water. These solids simply provide a coalescence surface and do not permanently hold the liquid. They simply allow the liquid to drain off in larger drops when the surface becomes covered.

## ELECTRICAL DEMULSIFICATION

The resolution of petroleum emulsion by electrical methods dates back to 1906. At that time Dr. Frederick G. Cottrell was a professor at the University of California. He had developed a technique for removing entrained particulate matter from gas streams employing an electrostatic field. This is the Cottrell Precipitator for which he is well known. He is not as well known for his work in the petroleum area. It is said that his attention was called to oil-field problems by some friends who had invested in an oil well that was producing the, then worthless, roily oil. He then applied his methods to oil-field emulsions, and in 1911 The Petroleum Rectifying Company of California was formed. Others have since built on the foundation he laid.

Let us ignore the patent literature for a minute. There are several papers discussing electrical dehydration of crude oil [15-19]. Dow [4] gives one of the first accounts. He states that the first commercial application of electrical dehydration was on a Lucile Oil Co. lease in the Coalinga field in California in 1909. This process used an alternating current of 11,000 volts [20].

The use of electrical methods to break emulsions is possible because most oil field emulsions are of the water-in-oil type, and thus, the system is relatively nonconducting. If the emulsion is water-external or if the oil is significantly conductive, electrical processes are usually not economical even if they are possible. Electrical emulsion-treaters do not operate on the principle of plating out the oil onto an electrode. When a drop of water moves into the electrical field, two separate actions take place. The drop is deformed by the charges induced on its surface, and it is attracted to other drops. The drops originally are nearly spherical with a radius r. When they have been deformed by the electric field, they assume the shape of an ellipsoid of revolution with semi minor axes a and b. In the absence of an electrical field a=b=r. It is possible to define a detailed function, F(a,b) which describes the shape of the drop in the field. It can be shown that

$$F(a,b) = E\left(\frac{r}{\sigma}\right)^{1/2} \tag{1}$$

where E represents the electric field, r is the initial radius of the drop, and $\sigma$ is the interfacial tension [16]. This distortion of the drop causes an increase in area of the drop and disrupts the interfacial film that is stabilizing the emulsion. Thus, the drops can coalesce, and the larger drops easily fall out of the oil.

In addition to distorting the drops the field also induces charges on the drops and thus generates a force of attraction between drops. This force can be evaluated from the equation

$$F = \xi_0 \xi E^2 \gamma^2 \psi\left(\frac{d}{r}\right) \tag{2}$$

where d is the distance between drops of radius r, $\xi_0$ is the dielectric constant, E is the field gradient, and $\psi(d/r)$ is a series function that is approximately equal to $6(r/d)^4$. A comparison of the electrical forces to the gravitational forces shows that in actual field conditions the electrical forces can exceed the gravitational forces by factors of 10 for large drops and even more for small drops.

The original Cottrell patent [20] describes a set of two concentric conical filter units that each consists of two layers of metal screen with a layer of filter cloth sandwiched between. It is specified that the cloth must be initially water wet. The two units constitute the electrodes and are connected to a source of electricity. The second claim specifies alternating current [20]. Three additional patents were issued on the same day, one to Cottrell and Cheever [21] and two to Cottrell and Speed [22,23]. All four patents were assigned to The Petroleum Rectifying Company of San Francisco, California. The Cottrell and Cheever patent eliminates the water wet cloth and provides means of rotating the inner electrode to "avoid short circuiting by limiting, retarding or breaking up the formation of chains of water globules." The first Cottrell and Speed patent, U.S. Patent 987,115 [22] emphasizes that the coalescing effect is not a cataphoretic effect and even implies that the use of direct current, wherein one would expect to get movement of the droplets toward one electrode, is of little help in resolving the emulsion. However, both alternating and direct current are claimed in the patent. The patent mentions the use of the direct injection of live steam to reduce the viscosity of the stream and also suggests the passage of the oil through the system at such a rate that the major part of the settling occurs in a holding tank subsequent to the passage through the electrode system. The other patent which was issued with it [23] concerns itself with electrode configurations so as to prevent shorting out of the system by chains of water droplets. In the early Cottrell-type treaters as described by Dow [4] there was a rotating or oscillating electrode as described in the Cottrell and Cheever patent [21].

Mr. Roy Lucas of the Petreco Division of Petrolite Corporation reports that as late as October of 1980 he saw a battery of eight units of these early electrical treaters in use in the

Comodoro Rivadavia field in Argentina. These units had electrodes which moved up and down about 8-10 in. They had been installed in 1923 and have been in use ever since. Later models went from the original, vertical configuration to a horizontal mode and the electrodes, although adjustable, were stationary when in use. High-voltage alternating current was used.

In 1919 a patent was issued to Seibert and Brady on an electrical treater using relatively low-voltage direct current [24]. The patent recommends that the process works best on concentrated emulsion and, therefore, it was designed to be used on the BS obtained from centrifuges or from mechanical separators. Dow mentions this process but does not elaborate [4].

Many modifications have been made to the electrical-dehydration equipment since that time. One of the next changes was to the Horizontal Field (H.F.) unit. Horizontal field refers to the configuration of the electrodes. This type was supplanted by the horizontal-concentrated-field (H.C.F.) and the concentric-ring (C.R.) types around 1925 to 1930. In these units turbulence was introduced into the oil stream to prevent chaining of the water drops without the mechanical complications of the moving electrodes. The H.C.F. type was said to be very good on tight emulsions and was used in batch treating. The C.R. type was said to be capable of much higher throughput than the earlier types.

Shea [5] in 1939 mentions three general types of electrical-dehydration plants in use in field demulsification, batch treating, flow-line plants and gravity-flow plants. He points out that electrical dehydration is most economical when it can be done on a large scale. Thus, it is often better to combine the output of several wells or several small fields to get most efficient results. On the other hand, it is better to run systems continuously and as closed systems so as to minimize stabilization times and evaporation losses.

Many configurations of electrodes have been patented. Worthington patented a system of horizontal, cylindrical electrodes, one of which rotated within the other [25]. Meredith discloses a propeller to direct the oil between plates and an electrode which rotates between the plates [26]. In 1921, Alden and Eddy received a patent on a system whereby the distance between the electrodes was varied by reciprocating action [27]. Cited against this patent but not prevailing was a patent by Harris [28] and two patents by Laird and Raney [29,30]. The Alden and Eddy patent also claims the process of varing the intensity of the electric field instead of moving the electrodes. Eddy also obtained three other patents [31-33] in which the emulsions are treated by other methods before being exposed to the electric field. McKibben patented a process where the oil was heated under pressure above the boiling point of water and then electrically treated [34]. On the same day he received a patent on a system where a long, narrow path between the electrodes and high, linear velocity of the oil were combined [35]. Fisher discloses the use of chemical treatment before using electricity [36]. Morrell teaches the addition of iron filings to the oil to aid breaking [37]. On the other hand, Eddy [38] has a process where the oil is treated under a partial vacuum. Lawrason combines acoustic, or vibratory, waves with electricity [39].

By this time it was becoming more difficult to come up with really novel variations. Herbsman [40] discloses details of an electric dehydrator, but the only claim that was allowed was the use of "chloramine toluene" as a demulsification reagent. Cage discloses two methods of relating the potential gradient to the conductivity of the oil [41]. It became increasingly difficult to maintain the high-voltage entrance bushings. Two patents cover the design of a "gravity-liquid heat barrier" which keeps the entrance bushings cool and isolated thermally from the main body of the treater [42, 43]. Another problem arises when attempts are made to dehydrate fuel on board an ocean-going vessel. The pitch and yaw of the vessel disrupt the settling process and upset level-maintaining devices. Watson and Jarvis received a patent in 1973 [44] on the design of an electrical treater which was compact and not affected by the motion of the ship. Other specific applications have been patented [45].

As late as 1976 a British patent was issued on an elaborate electrode configuration [46]. Woelfin, in 1958, applied for a patent in Germany on a concentric configuration. The patent was issued in 1972 [47]. A 1966 Russian patent deals with another special electrode configuration [48] as do five other German patents [49-53].

In all, several hundred patents have been issued in the area of electrical dehydration. Some of them are hard to identify or locate since the titles do not always contain key words, and patents on insulators or distributors for the electrical systems may not appear in the same patent section. However, for most purposes it is not necessary to carefully study the patent literature. Anyone who feels that a particular emulsion problem could be solved by electrical treatment can get plenty of technical help from the manufacturers of these units. Advertisements for these companies can be found in the trade journals [54-57].

## DESALTING

Today, while electrical dehydration is still important as a method of field dehydration, the really large applications are in the refineries where the crude is "desalted" by deliberately adding water to dilute the brine or dissolved solid salts. The water is then removed in electrical units. Spent acid and caustic are also removed from refinery streams by electrical treaters. In addition, electrical units are being used to clean up residual fuels for use in power turbines [58,59].

## CHEMICAL DEMULSIFICATION

The most widely used method of treating crude-oil emulsions is chemical demulsification. Much has been written on this subject. The patent literature is particularly voluminous. A search of the patent literature will turn up between 2,000 and 3,000 patents directly referring to this subject and thousands more related to it.

References 3-10 at the end of this chapter give general discussions of the chemical treatment of petroleum emulsions. The process is simple in principle. A material that will cause the disruption of the emulsion and the coalescence of the dispersed droplets is added to the crude emulsion and mixed thoroughly. The reagent is given some time to take effect during which time gentle mixing may be of help. The oil is then held in a quiescent state long enough to allow the water to settle out. It is obvious that any of the mechanical arrangements described earlier in this chapter can be used in conjunction with chemical treatment.

Dow, on page 45 of Ref. 4 lists six classes of chemicals that can break emulsions:

1. Compounds that can take up water and react with it. Under this classification he mentions calcium chloride, quicklime, and Portland cement.

2. Compounds that cause flocculation of the substance making up the protective film. Here he lists sodium chloride, ferrous sulfate, and sodium sulfate.

3. Compounds that can react with salts in the water or with any organic acid that may be present. These compounds include lye, sodium bicarbonate, and calcium oxide.

4. Compounds that tend to form emulsions of the opposite kind. Here Dow lists iron oxide, finely divided silica sand, clays, sodium soaps, resin and resin soaps, albuminous materials, gums, starch, and organic sulfonates.

5. Electrolytes that can neutralize charges. He includes sulfuric acid, acetic acid, ferric chloride, and ferric nitrate.

6. Solvents for the protective film. Carbon disulfide, benzene, acetone, alcohol, ether, and carbon tetrachloride are among the materials mentioned.

It is clear that almost anything that was handy was tried as an emulsion breaker and that, at least in some cases, almost anything seems to have worked. Dodd [60] suggested the use of phenol and sulfuric acid. Barnickel patented the use of copperas [61]. He later patented the use of water-softening agents [62,63]. These agents were marketed under the trade name Tret-o-lite. One of the first materials was essentially tincture of green soap. Another patent covered the use of "Turkey red oil" [64]. The William S. Barnickel Company merged with the Petroleum Rectifying Company in November 1930 to form The Petrolite Corporation. Since that time a steady stream of patents has emerged from the research departments of that and other companies marketing emulsion-breaking chemicals. Melvin DeGroote of the Tret-o-lite Division of the Petrolite Corporation had his name on over a thousand patents before his retirement.

The methods of adding chemical to the crude-oil stream, were at first, very simple. Early pictures show a can containing chemical hanging in the manhole of a storage tank. A nail hole in the bottom of the can allows chemical to drip into the tank. Incoming oil mixed the chemical, and the separate water is drawn off the bottom of the tank. Once a day the "pumper" refilled the can. It was soon found that the point of addition and the method of addition of the chemical were important in attaining success. In discussing the flocculation of oil-in-water emulsions we noted that the process was accomplished in three stages. First, the chemical was mixed in well; then the floc was allowed to form with gentle stirring; finally, the floc was allowed to settle without stirring. Much the same concepts apply to chemical treating of crude-oil, W/O emulsions. The chemical is introduced into the system in such a way as to insure good mixing. It is then given time to take effect before passing the oil to a settling zone. Exactly how these steps are achieved depends on the condition of the oil and the configuration of the producing equipment.

## BOTTLE TESTS

A better understanding of the process can be obtained by studying the method recommended for testing de-emulsifying chemicals. This method is described in detail in pages 33-44 of Ref. 3. To run the test some six-ounce, screw-top prescription-bottles graduated to 100 ml, a water bath with thermostatic control, solutions of de-mulsifying compounds, and some pipettes or syringes are needed. A few breakers and graduated cylinders are also useful.

After obtaining a fresh, representative, sample of the emulsion, the test is run as follows:

1. Fill several bottles with emulsion to the 100 ml mark.
2. Select a compound to be tested and add progressively larger amounts of chemical to a series of bottles. The compound is preferable added as a solution in some appropriate solvent. Note!! Do Not Pipette by Mouth as Shown on Page 39 of the Above Reference.
3. Label the bottles so that you can identify them later. Cap them securely.
4. Place the bottles in a shaking machine or shake by hand for about 100-200 shakes. In hand shaking the bottle necks are held between the fingers and, with three bottles in each hand, the arm is moved back and forth about 12 in. at about 100 shakes 1 min.
5. Observe any changes in appearance. Most oil-field emulsions will be sort of coffee-with-cream color as received. If the emulsion is breaking, the color will get darker, and drops of water will appear on the sides of the bottle. If no changes occur, another compound must be tried. If separation does occur, then further tests should be run to determine the amount of mixing that may be required, the optimum amount of chemical, and the expected time that will be required to get separation. The tests should be conducted at the temperatures that are expected in the field system. Breaking must be fast enough to be completed in the residence time of the system.
6. When the optimum agent and amount have been established, a BS & W test should be run on the recovered oil to see if it meets specifications.

A certain amount of experience is required to learn to match the bottle-test results with actual field performance. There are apparently some subtle differences in shaking and addition techniques that can alter the results of the tests. There are certain oil fields where emulsions have been successfully treated with particular combinations of chemicals for years, yet a bottle test would show the chemical to be ineffective. Fortunately, these cases are rare. Often, however, the emulsion will not break completely, and extensive tests are required to find a material that will do the job. This is where the experience of a seasoned field-tester is required. The art of emulsion treatment comes in at this point. Most oil-field operators avail themselves of the services of the chemical companies' field agents who have experience in that field and have a library of materials to try.

Several precautions must be taken to insure reasonable results in demulsification tests. The emulsion must be representative of the real field condition. Remixing a partially separated emulsion before testing may change the whole picture. Emulsions from certain fields can be sampled and shipped all over the country yet still be stable and respond to testing in a realistic manner. Other emulsions have to be tested at the wellhead or actual field tests must be run. Any laboratory tests must be field-verified, but gradually, a good field-tester gets a feel for what is needed.

By closely observing the changes in the bottle test it is sometimes possible to get a good idea of the optimum amount of mixing required or the induction period that may be needed. One of the most common errors of a beginner in the field is to assume that if a small amount is good, a larger amount is better. This does not necessarily follow. It is just as possible to use too much chemical as too little. Some materials and some emulsions are very easily overtreated; other combinations are not. There are several general demulsification compounds that will work on a wide variety of emulsions without overtreating. They are usually needed at much higher concentrations than the best compounds, but they can be "slugged" to a system without too much risk. These materials are sometimes called "knockout drops" and are used in testing to determine the total amount of BS & W that can be removed from a sample.

## TREATMENT METHODS

Once a material has been found that will break the emulsion, it is still necessary to determine how and where the material should be introduced. In some cases the compound is diluted with lease oil, and batch treated down the production string of the well. A more common method of downhole treatment is to introduce the material into the annular space between the casing and the tubing. The chemical can be dispersed into water and washed down the well. It can be added continuously as a water dispersion or as a solution in distillate or lease curde. In hydraulic pump systems the chemical can be added continuously to the power oil.

If the chemical is not added downhole, it can be added to the gathering manifold at the wellhead or to the tank battery. Usually, it is preferred to add the material before the free-water knockouts or the gas separators. The injection pump for the chemical is designed to pump against whatever pressure is in the lines. This pressure can sometimes be quite high. Injection pumps can be driven by electricity or by gas pressure, or they can be coupled to the walking beam of the well pump.

If there are transfer pumps in the gathering system, the chemical may be injected just ahead of the pump to help insure good mixing. The point of injection should be chosen so that good mixing occurs, and so that the induction period for the demulsification process will be completed at a point where the water can be separated. If the water drops out too soon, it may be reemulsified by pumps or valves. If the allowed induction period is too short, the heater treater or other equipment may not be used to best advantage. If electrical demulsification is being used, it is often supplemented with chemical treatment. Electrical treaters are most efficient if they can be run at full capacity and in the larger sizes. For this reason the oil from several leases may be combined before being run to the treater. If production from each of the leases is variable, the composition of the mixed feed may vary. This may make it difficult to select the best chemical. Some combinations that work in the lease separators may not work well on the combined stream.

Originally, the first oil fields were broken up into a large number of small leases. Each lease was owned and operated by a different entity, and each was responsible for producing salable oil. Gradually, the older oil fields were unitized, and a single authority was established to produce the field as a unit. This system is almost a must in the case of secondary or tertiary recovery operations. As the fields become unitized, the demulsification tends to be done in large, central treatment facilities. If oil is produced offshore or underwater, several wells may produce to a single platform. The oil may be treated at the platforms or transferred in pipelines to a shore battery. These large, central treatment facilities can be mechanized and put on automatic control. The water content of the incoming oil can be sensed, and the quality of the treated oil continuously monitored. Lease automatic custody transfer (LACT) units test the oil, meter it into the pipeline, and credit the owner an amount based on the amount of BS & W and the API gravity of the crude. A central facility of this type has been known to handle 80,000 barrels (a barrel is 42 gal) of crude a day.

Early in the history of oil production it was recognized that oil was not like other ore bodies. A mineral deposit is fixed in place and belongs to the owner of the land under which it occurs. On the other hand, the first wells in an oil field can pull oil from under someone else's property. Thus, as soon as a discovery well was brought in, all leaseholders made a mad rush to get as many wells drilled on their property as possible and to produce these wells as fast as possible. This often resulted in the loss of gas pressure, intrusion of water, and bypassing of considerable amounts of oil. To combat these problems, laws were written in various states controlling the spacing of wells and the amount of production allowed from each well. The prorationing of the production usually took the form of running the well for a certain number of days a month or until a certain amount of oil was produced. If the quota was not met in a given month, it was lost. In places where prorationing is still in effect it is common practice to produce for a certain number of days at the end of one month and at the beginning of the following month, then to shut the well in until the end of the following month. This approach concentrates production and promotes economical operation, but if anything goes wrong, production may be lost forever. Anything that interrupts the production of the oil must be remedied as soon as possible. This fact has tended to discourage do-it-yourself demulsifer selection.

If an emulsion problem crops up, the lease operator wants immediate help. All major venders of demulsification chemicals and equipment keep a staff of specialists permanently stationed near all major oil-production sites. The larger oil producers also have their own treating experts. These people are on call at all times and are prepared to do all they can to get the units running and keep them on line. Of necessity, these people tend to be highly pragmatic. Theory may be nice, but if something works, use it. We would all like to be able to feed certain figures into a computer and get a printout that would tell us what to use, and how much to use, and how to apply it. This must be considered a utopia at the present time. Thus, the problem must be still considered to be an art as well as a science.

## REVERSE EMULSIONS

So far we have been talking about water-in-oil emulsions. What about reverse emulsions? This brings us back to many of the principles discussed in Chapter 3. In the language of the oil fields, reverse emulsions are oil-in-water emulsions. They are nowhere nearly as common as the water-in-oil type. They do occur, however, regularly in certain places and instances. In some of the California and Rocky Mountain oil fields much more water is produced than oil. These fields tend to produce reverse emulsions. The connate water is usually fresh water as opposed to brines. When they do occur, the reverse emulsions are often very stable. The production looks like muddy river water. The oil does not separate on standing. Heat can be used to break these emulsions, and in the case of thick, low-gravity crudes it has to be used, but the cost of heating all the water is significant. Heat exchangers can be used. Even with heat, a treating chemical is usually needed. Adjustments of pH sometimes help, and inorganic salts have been used. In many cases the problem is not amenable to a simple solution. When the production fluids reach the surface, they are found to contain some free oil and some reverse emulsion. At the free-water knockouts we get some oil with water emulsified in it, and the water that settles out contains oil. The oil stream is treated by standard methods. The water is treated with a "reverse compound" and allowed to settle, or may be sent to a flocculator to separate the oil. The oily sludge that is recovered from the flocculator is then further treated and sent back into the oil stream. If the flocculator sludge contains too many solids, it may be necessary to wash the sludge before running it into the oil stream.

Several conditions combine to complicate the treatment of reverse emulsions. First, they usually occur where the connate water is fresh, so the density difference between the oil and the water is small. Settling is thus slowed. Sometimes the oil is even heavier than the water and settles to the bottom. This case requires special equipment. Secondly, the reverse emulsions often occur in secondary or tertiary recovery projects. Here, besides the large amounts of water produced, we have to contend with the detergents that were used in the flooding process. These detergents are good emulsifiers (or they wouldn't be effective) and are present in greater concentrations than the naturally occurring materials in normal crude. Furthermore, they are often ionic in nature and almost require stoichiometric treatment. Treating rates for these systems are an order of magnitude higher than for normal systems. Besides the detergents used in the flood there are also the polymers used to control the flood front, and there may be solids that were carried along. Enormous amounts of chemicals are used in a secondary recovery project, so naturally, there is a need to use the least expensive detergents available. One type of detergent that is fairly cheap and readily available to the petroleum industry is the sulphonates that are the by-product of the refining of kerosene. Feed stocks in this range are often treated with sulfuric acid to remove unsaturates. The result of the reaction is the production of considerable amounts of petroleum sulphonates. These compounds are a captive product and relatively cheap. They make good waterflood detergents, but when they begin to be produced back with the oil at the end of the flood, they present a problem. They make very stable emulsions, usually of the reverse type, and they must be removed from the oil before it can be refined. If they get through the refinery and show up in the asphalt or residual oil, they cause more problems. They raise the sulfur content of the fuel, contribute to acids in the stack gases, and lower the melting point of the asphalts. It begins to look as though it will be necessary to select the detergents used in the floods so that they can be deactivated at the other end and either recovered or destroyed inexpensively.

Two other tertiary recovery methods give rise to reverse emulsions--fire—floods and the steam floods. In a fireflood, the oil-bearing formation is set on fire, and a controlled combustion is maintained by the injection of air down some wells. The heat of combustion fluidizes the oil ahead of the fire front and drives it to production wells. The heat also cracks the oil and produces light ends and gases. The produced fluids are hot and under pressure, and superheated steam and gases are present. Some really tough emulsions can result. Most of them have been successfully broken, but doing so is not usually cheap. In the steamfloods two different systems are used. The steam can be injected and used to drive the oil to a production well or the huff and puff system can be sued. The straight steam displacement has not been too successful because of the vast amount of rock that has to be heated. The huff and puff system has shown better results. In this system the steam is forced down a well for a period of time. The well is then allowed to flow until the pressure equalizes again. Large amounts of water and condensate are produced. Some very stable reverse emulsions result. Some success has been claimed by downhole injection of emulsion preventers. Their effectiveness is, however, erratic, and further treatment is usually required at the surface.

## SUMMARY

How, then, do we summarize the demulsification problems of the petroleum industry? After 70 years they have settled down to a reasonable clear picture from a performance basis. Where possible, conditions are controlled to minimize emulsion formation although not all opportunities have been realized in this area. The emulsions that do occur are, by and large, dealt with efficiently. Mechanical and electrical methods are usually supplemented by the use of chemicals. Chemical treatment seems to account for most of the field demulsification. Vendors of demulsification chemicals are able to develop chemicals that will treat most emulsions, and the efficiency of these products has increased enormously over the years. Even with recent inflation, the cost per barrel to treat most emulsions by the best methods is still lower than it was years ago. The emulsions produced by recent recovery methods are being broken in most cases.

So much for the practical side. The theoretical side is something else. We know, in general, what causes emulsion and how they are stabilized. In particular cases, we do not know much more than we did 50 years ago. It is known that some materials in the crude are natural emulsifiers. The compounds in certain crudes have been isolated and identified. The problem is that there is so much variability among crudes and among oilfield conditions that generalizations, if true, are not much practical help. There is a large store of practical information about what works in certain cases. Most of this information is anecdotal and fragmentary. Even where there are detailed and careful measurements, it is difficult to isolate the relevant material. In a sense we have too much data. There are too many variables, and it is not known which are relevant. Just because a material works in a given place doesn't mean it is the optimum formulation. Just because it works today does not mean it will be the best choice tomorrow. Enough has been learned to get the job done, but we don't always know just what the detailed interactions at the interface are.

With more and more concern with toxicity and pollution, all industry will have to deal with emulsions. By and large the oil industry has done a good job. Their experience can be of great help in solving other problems.

One might well ask, "Is anything done about our lack of knowledge in the area of demulsification theory?' You will get several different answers. The textbooks leave the impression that we know all there is to know about the subject. We have Stokes' law to handle settling, Einstein's equations to cover Brownian movement, the Deriaguin, Landau, Verwey, Overbeck (DLVO) theory, the Helmholz double layer, Hydrophillic-lipophilic balance (HLB) and on and on. It is clear that emulsions are stabilized either by charge repulsion or an interfacial film. All you have to do is adjust the zeta potential or disrupt the film, and your troubles are over.

Why isn't someone working on these problems? To be fair, it is not an easy area to investigate rigorously. One can use the Edisonian approach —test many formulations and come up with specific answers that will solve specific problems and can be sold at a profit. Often it seems that there is time for no more than this. But some work is being done. It is just that it is very difficult to work out techniques that will give valid results. Determining the phase diagram of a single system may very well take over a year. Thousands of observations must be made on, at least, hundreds of specific mixtures. The temperature and pressure must be controlled. Phases must be separated, analyses run. It is long and tedious. Few graduate students care for such a task. Few companies feel that it is worth the time. However, some studies are underway.

We will look at some of the recent work and see what is being done. A good example of the state of the art is a patent that was issued on January 16,1979 [65]. The patent is titled, "Method for Selecting a Demulsifier for breaking a Water-in-Oil Emulsion". The invention actually turns out to be a device for determining whether a selected demulsifier is effective on a particular emulsion. The method of selecting the demulsifier to be tested is not given. Rather, ". . . a demulsifier is selected. . . the foregoing steps are repeated for a plurality of selected demulsifiers . . . " A tidy description of Edisonian screening. No clue is given as to how to select what to test.

Several specific articles will be cited to show the difficulties inherent in obtaining meaningful results in this area of investigation.

Consider a paper titled "Stabilization of Oil-in-Water Emulsions by Nonionic Detergents V" [66]. This is one of a series of papers by the same authors covering a lot of detailed work. Measurements were made of densities, interfacial viscosity, electrophoretic mobility, and zeta potential. The paper represents a large amount of careful work yet presents no conclusive results. They conclude,

"An attempt was made to interpret the results in terms of the DLVO theory and the theory of 'entropic' stabilization. The results reinforce previous data which indicated that entropic forces are the main stabilizing forces in these systems and that the reduction in stability on the addition of the salt is due to changes in the nonionic film which are too subtle to detect by interfacial tension or viscosity measurements."

Consider now a paper on "The Relevance of 'HLB' to De-Emulsification of a Mixture of Heavy Oil, Water and Clay" [67]. The oil used is an actual oil that is being produced in Canada. The surfactants tested are materials that can be bought on the open market in commercial amounts. "The present work considers the possibility of correlating the de-emulsifying ability of several series of synthetic sufactants with their Hydrophilic-lipophilic balance (HLB), an easily measured parameter. . . The HLB values were obtained from literature supplied by these companies." That is the only easy way to measure HLB. The determination of the HLB of a completely new material is not an easy matter. It is a long, drawn-out procedure that in the end, measures what mixture of known materials most resembles the new material. Calculations of the HBL can give several results depending on the method used. Likewise, the paper represents a large amount of careful work with no conclusive interpretation. "The most effective agents for dewatering had HLB values between either 4 and 6 or 13 and 15." The concentration of demulsifier that was used in the test was 1.5 g of reagent to 50 g of emulsion, and the emulsion was cut with an equal amount of toluene before testing. The emulsion was shaken for 1 hr and allowed to settle for 3 hr. These are not realistic concentrations or conditions for field applications.

There are several emulsion parameters which can be measured but which seldom seem to yield correlatable results. Whenever two liquids are in contact, an interfacial tension can be said to exist between them. It would seem logical to assume that there should be some relationship between the interfacial tension and the stability of the emulsion. Any material added to a system could be expected to change the interfacial tension, and the HLB could be expected to be some measure of this effect. Never forget that a statistical correlation between two factors does not necessarily imply a causal relationship.

Let us now review an excellent paper even though it is nearly 15 years old. It is called, "Effect of pH on Interfacial Films and Stability of Crude Oil-Water Emulsions" [68]. In this work, 45 crude oils from 21 reservoirs throughout the world were tested. The amounts of reactants and the test conditions were realistic. In addition to the measurement of emulsion stability, the results of another test that bears study are presented. This test is called the "film-ratio" determination. In this test, oil is allowed to flow from the tip of a capillary into an aqueous bulk phase. Just enough oil is extruded to form a pendant drop. The drop is allowed to age for a specified time; then the oil is withdrawn into the tube until the surface of the drop starts to wrinkle. Enough oil is returned to the drop to remove the wrinkles, and the drop is photographed both initially and at the end of the test. It is possible to calculate the area of the drop at both stages. The ratio of the cross-sectional area of the compressed drop to the original drop is the film ratio.

In the film ratio test, crude oils seem to fall into three general classes: those that form solid or rigid films; those that form films that readsorb and disappear; those that do not form visible films. A crude oil may vary in its film-forming properties according to the aqueous phase it is tested against. The conclusions from this paper are quoted verbatim [68].

1. Emulsions commonly encountered in producing crude oil are stabilized by films that form at the water-oil interfaces and interfere with coalescence of dispersed water droplets. These emulsion-stabilizing films are composed of interfacially active materials from the asphaltene and resin fractions of the crude oil combined with ions from the aqueous phase and insoluble fines in the crude oil-brine system.

2. Emulsion stability can be decreased by conditions that reduce the film-forming capacity of the crude oil. These films can be minimized or their physical properties altered by treatment with acids and bases alone or in conjunction with inorganic salts, surfactants, heat or electricity.

3. An optimum pH, at which value maximum emulsion breaking occurs, exists for most crude oil-brine systems. The optimum pH for maximum emulsion instability depends on both crude oil and brine composition. Brine composition is the more important from the standpoint of economics of pH adjustment.

4. Laboratory results indicate that many crude oil emulsions can be demulsified at significantly lower costs if they are treated at their optimum pH values. It is not common field practice at this time to adjust emulsion pH.

5. In addition to pH, many other factors such as crude oil viscosity, water-oil ratio, brine composition, fine solids, agitation, temperature and aging also affect emulsion stability. Because of these numerous variables, the final optimum emulsion-treating system must be established by field tests. Laboratory tests will serve as useful guides to the field tests.

Conclusions 1 and 2 are reasonable and in line with our previous findings. Conclusions 3 and 4 show that the pH of the emulsion (presumably the aqueous phase) does affect the stability of the emulsion and that the most efficient demulsification can be done at some optimum pH. However, note the last comment: "Because of these numerous variables, the final optimum emulsion-treating system must be established by field tests." The author again emphasizes the need of empirical tests.

In this paper it is stated that the materials that act as emulsifying agents are "interfacially active materials from the asphaltene and resin fractions of the crude oil" [68]. Strassner also states that the asphaltenes form the most rigid films at low pH, while the resins form mobile films that are strongest at high pH. While this paper does not give an easy way to select demulsifiers, it does give valuable insight into the factors producing emulsions in crude-oil systems.

Another paper which gives further insight is "Formation, prevention, and Breaking of Sea Water-in-Crude Oil Emulsions 'Chocolate Mousses' " [69]. This paper discusses the formation of emulsions in oil spills. It seems that the floating oil will take up drops of water from wave action and a concentrated water-in-oil emulsion will form. These emulsions look like a chocolate mousse. They can contain as much as 70% seawater and are very stable and very viscous. The authors report that the mousses form most readily at or near the pour point of the oil. If the oil has been dewaxed, it does not produce mousse. They also report that the physical state of the wax affects the mousse formation. Different heating and cooling sequences can change the pour point and the emulsification tendencies. If a pour point depressant is used to lower the pour point below the seawater temperature, mousse will not form. The presence of wax alone is not enough to cause emulsification. Asphaltenes also have to be present. Without the wax and the asphaltenes, the oil does not emulsify. Without either the wax or the asphaltenes, the oil does not emulsify. With both the wax and the asphaltenes, the oil forms mousse. With the wax and only 10% of the asphaltenes, the oil also forms mousse. If the mixture of wax and asphaltenes is added to a refined lube-oil, it also forms mousse. If attempts are made to form mousse at elevated temperatures above the pour point of the crude, they do not succeed. But, if a mousse is made at ambient temperature and then heated, it does not break. The authors, ". . . presume that . . . when the wax is redissolved, the asphaltenes remain at the interface and . . . preserve stability." The authors then describe their tests on a breaker for the mousse called LA 1843. This material is not further described. Again, we have some interesting data on the complexity of the problems but no easy answer.

Finally, let us look at a typical older paper [70]. This paper is titled, "Some Basic Concepts Regarding the Separation of Oily Water Mixtures." This paper refers to Stokes' law and the effect of interfacial tension on the work required to form drops. It states that for very small drops it will be necessary to use mechanical separators rather than simple settling. The pertinence of viscosity as a parameter is mentioned. Methods to calculate conditions for breakthrough of the continuous phase are given. The next section states, "Capability of Mechanical Separators is Best Determined by actual Test Data." Mechanics of coalescence are discussed with diagrams and pictures, and the possibility of chemical treatment is discussed. The author concludes, "Unfortunately, the selection of an effective counter-surfactant to promote coalescence is based on empirical testing." The author is absolutely correct.

Citing these specific articles has reinforced our previous contention that this is an area where there are so many variables, and where the techniques for investigating the fundamental problems are so tedious or difficult, that it is easier on a practical basis to rely on empirical field testing. This approach becomes an art and, at least for now, may be the only route. However, as we increase our basic knowledge of the effects of various factors on the stability of petroleum emulsions, it is hoped that eventually we will be able to select demulsifiers on a more scientific basis.

## REFERENCES

1. *Thomas Register*, Thomas Publishing Co., One Penn Plaza, New York (1982)
2. *Hydrocarbon Processing Catalogue*. Gulf Publishing Co., Box 2608, Houston, Texas 77001 (1981)
3. *Treating Oil Field Emulsions*, Petroleum Extension Service, University of Texas, Austin, Texas, and American Petroleum Institute, 1st Ed. 1949, 2nd Ed. 1955, 3rd Ed. 1974, 4th Impression 1978

4. D. B. Dow, *Oil Field Emulsions*, U.S. Dept. of Commerce Bulletin 250, 1926
5. G. B. Shea, *Practices and Methods of Preventing and Treating Crude-Oil Emuslions*, U.S. Dept. of the Interior, Bureau of Mines, Bulletin 417, 1939
6. L. G. Uran, *National Petroleum (News)*:51-57 March 27,1929
7. F. Steinhauff, *Petroleum (London)* *25*:294-297 and 335-337 (1962)
8. P. L. Bansbach, *Oil and Gas J.*, *52*:87-93 (1970)
9. J. L. Wiggins, *Pet.-Eng.*, *29*:B47-B56 (1957)
10. C. M. Blair, Jr., *Oil and Gas J.* *44*:11 (1945)
11. D. B. Dow, *Oil Field Emulsions*, U.S. Dept of Commerce Bulletin 250, 1926, p. 257
12. H. F. Fisher, U.S. Patent 1,932,093 (1933)
13. G. H. Meinzer, U.S. Patent 1,911,839 (1933)
14. G. H. Meinzev. U.S. Patent 1,912,840 (1933)
15. D. E. Jackson and J. C. Scott, *Petro. Chem. Engr.* *33*: 234 (1961)
16. L. C. Waterman and S. D. Winslow, *IIEE Convention Record* pp. 9-22 (1966) preprint.
17. R. N. Lucas and R. W. Greenlee, *Proceedings of the Annual Meeting of the American Society of Lubrication Engineers*, (1970) preprint 36 pp.
18. R. W. Greenlee and R. N. Lucas, *ASTM Special Technical Publication* *531*:73-104 (1973)
19. W. G. Eddy and A. C. Eddy, *Ind. Eng. Chem.* *13*:1016 (1921)
20. F. G. Cottrell, U.S. Patent 987,114 (1911)
21. F. G. Cottrell, and A. Cheever, U.S. Patent 987,117 (1911)
22. A. G. Cottrell and J. B. Speed, U.S. Patent 987,115 (1911)
23. A. G. Cottrell and J. B. Speed, U.S. Patent 987,116 (1911)
24. F. M. Seibert and J. D. Brady, U.S. Patent 1,290,269 (1919)
25. J. T. Worthington, U.S. Patent 1,783,595 (1980)
26. W. Meredith, U.S. Patent 1,452,207 (1923)
27. S. F. Alden and A. C. Eddy, U.S. Patent 1,394,462 (1921)
28. F. W. Harris, U.S. Patent 1,281,952 (1918)
29. R. E. Laird and J. H. Raney, U.S. Patent 1,116,299 (1914)
30. R. E. Laird and J. H. Raney, U.S. Patent 1,142,759 (1915)
31. A. C. Eddy, U.S. Patent 1,826,267 (1931)
32. A. C. Eddy, U.S. Patent 1,796,750 (1931)
33. A. C. Eddy, U.S. Patent 1,779,795
34. C. W. Mckibben, U.S. Patent 1,299,590 (1919)
35. C. W. Mckibben, U.S. Patent 1,299,589 (1919)
36. H. F. Fisher, U.S. Patent 1,838,379 (1931)
37. J. C. Morrell, U.S. Patent 1,837,714 (1931)

38. A. C. Eddy U.S. Patent 1,847,602 (1932)
39. L. Lawrason, U.S. Patent 1,942,480 (1934)
40. A. M. Herbsman, U.S. Patent 1,931,112
41. J. M. Cage, U.S. Patent 1,754,009 (1930)
42. D. W. Turner, U.S. Patent 3,719,584 (1973)
43. F. D. Watson and J. D. Wristow, U.S. Patent 3,926,774 (1975)
44. F. D. Watson and A. R. Jarvis, U.S. Patent 3,736,245 (1973)
45. F. D. Watson, U.S. Patent 3,794,578 (1976)
46. Howe-Baker Engineers Inc., Brit. Patent 1,437,682 (1976)
47. W. Woelfin, Ger. Patent 1,420,894 (1972)
48. I. N. Rybachok, U.S.S.R. Patent 183,313 (1963)
49. D. N. Turner, Ger. Patent 1,261,825 (1959)
50. E. L. Shirly and L. Waterman, Ger. Patent 1,215,651 (1966)
51. O. Gupner, Ger. Patent 1,179,912 (1964)
52. H. Stock, Ger. Patent 1,014,076 (1964)
53. G. Grave, Ger. Patent 947,414 (1956)
54. *Hydrocarbon Processing*, Gulf Publishing Co., Houston Tex.
55. *Oil and Gas J.*, Penn Well Publishing Co., Tulsa, OK
56. *Oil Week*, McMan Hunter, Calgary, Alberta, CAN
57. *Erdsel Erdgas Zeitschrift*, Urban-Verlag-Hamburg, W. GER
58. R. W. Greenlee and R. N. Lucas, Paper Presented at E.P.R.I. Combustion Turbine Residual Fuel Workshop, July 1980, Atlanta, GA preprint
59. R. W. Greenlee and R. N. Lucas, *ASTM Special Technical Publication 531*: (1973)
60. A. V. Dodd, *Chem. and Met. Eng. 28*:249-253 (1923)
61. W. S. Barnickel, U.S. Patent 1,093,098 (1914)
62. W. S. Barnickel, U.S. Patent 1,223, 659 (1917)
63. W. S. Barnickel, U.S. Patent 1,467,831 (1923)
64. M. Degroote, U.S. Patent 1,596,596 (1926)
65. W. B. Allen, J. W. Harrel, and W. W. Webster, U.S. Patent 4,134,799 (1979)
66. P. H. Elworthy, A. T. Florence, and J. A. Rogers, *J. Colloid and Interface Sci. 35*:23-32,34ff.(1971)
67. D. G. Cooper, J. E. Zajic, E. J. Carmel, and J. W. Wood, *Can. J. of Chem. Engin. 58*:576-579 (1980)
68. J. E. Strassner, *J. Pet. Tech. 20*:303-312 (1968)
69. A. L. Bridie, T. H. Wanders. W. Zegveld, and H. B. Van Der Heijde, *Marine Pollution Bulletin 11*:343-348 (1980)
70. G. P. Canevoir, *ASLE Transactions 12*:190-198 (1969)

# 6

# Additional Methods and Areas in Demulsification

## INTRODUCTION

Another way of classifying emulsions is to divide them into six classes according to the type of external phase and the phase ratio. The six classes are:

1. Oil-in-Water
   a. 1-30% internal phase (low)
   b. 30-75% internal phase (medium)
   c. >75% internal phase (high)
2. Water-in-oil
   a. 1-30% internal phase (low)
   b. 30-75% internal phase (medium)
   c. >75% internal phase (high)

The three classes that are the most difficult to deal with in industry are 1.a, 2.a, and the combined classes of 2.b and 2.c. The first class is the low internal-phase ratio oil-in-water emulsions; the second is the low internal-phase ratio water-in-oil emulsions; the third is the medium to high internal-phase ratio water-in-oil emulsions. The first class is encountered in many waste waters and effluent streams. The second is found widely in the petroleum industry. The third is found in the sludges from certain recovery processes. The oil field emulsions have been covered in Chapter 5, but the first and third classes will bear further study. This chapter will cover certain, less common, ways of dealing with dilute oil-in-water emulsions and of handling sludges.

One of the reasons that the dilute oil-in-water emulsions are difficult to resolve is that the individual drops are so small and so far apart that they will neither settle nor coalesce. If they can be crowded together in some way, they may be easier to recover. Crowding can be accomplished by a number of fairly new techniques. Among these techniques are electrodialysis, reverse osmosis, ultrafiltration, and filtration coalescence. Electrodialysis, reverse osmosis, and ultrafiltration are closely related. In electrodialysis the inorganic or small, organic ions are impelled through selective membranes leaving the effluent less concentrated in dissolved solids. This change in ionic strength, and perhaps in pH, will sometimes destabilize the emulsion and allow separation. In reverse osmosis, water molecules are driven through a semipermeable membrane, and the effluent is thus concentrated so that the oil drops are closer together and the ionic strength is increased. Both these changes can help to destabilize the emulsion. In ultrafiltration, water and other small molecules are removed through a membrane and, again, the more concentrated emulsion may be less stable.

Each of these approaches is inherently more expensive than conventional methods. Thus, for them to be used they must either accomplish what cannot be done by another method, or they must show other advantages, which is sometimes the case. Each one will be discussed in more detail.

## ELECTRODIALYSIS

Electrodialysis (ED) depends on two factors: first, the fact that if two electrodes are placed in an aqueous solution and a direct current passed through it, the ions in solution will migrate to the electrode of opposite charge; secondly, the fact that we can construct membranes that will selectively pass only positive or negative ions. Thus, if we set up a cell as pictured in Figure 6.1, we can attract the cations through one membrane and the anions through another membrane, combine these two streams, and get two effluent streams—one from which most of the ionic species have been removed and another in which they are concentrated. Any suspended solids and oil stay with the dilute stream. These cells can be stacked in parallel and operated continuously at substantial throughputs. Care must be taken to keep suspended solids or oils from plugging or fouling the membranes. Adjustment of pH may be necessary to avoid scaling.

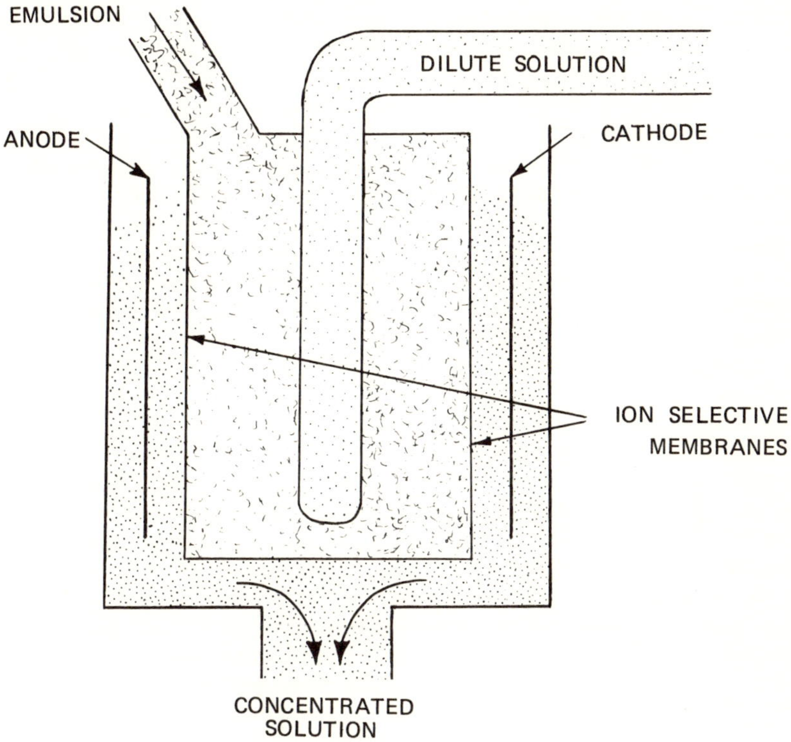


Figure 6.1 Electrodialysis cell.

Several secondary advantages can be realized by electrodialysis. In cases where the dissolved salts are worth recovering, the concentration of these materials into a stream without other solids or oil allows them to be evaporated and recovered. It is best to remove suspended solids from the stream before the ED cells if possible, and it is actually advantageous to have a stable oil emulsion so that the oil will not come out on the membranes. If the oil is then flocculated and removed, the effluent water may be clean enough to recycle. In some cases the oily water can be reused in other processes before oil recovery. Often the value of the recovered salts and the decreased overall use of water offset the cost of operating the cells. An excellent discussion of electrodialysis can be found in the book by Ouellette, King, and Cheremisinoff [1]. It is worth noting that electrodialysis of an emulsion does not concentrate the emulsion. It removes ionic species from the external phase. Their removal may help destabilize the emulsion.

We now look at a series of processes that are similar but different enough to be useful in different ways. In each of the processes a pressure difference is established across a "membrane." The nature of the membrane determines the type of process. In electrodialysis the ions are driven by an electric field. In these processes the materials are impelled by a pressure difference. Fundamentally, the processes differ in the size of the particles that can pass through the membrane. The membranes are neutral and do not select on a charge basis.

## REVERSE OSMOSIS

In ordinary osmosis water diffuses through a semipermeable membrane from a dilute to a more concentrated solution. The overall effect is to dilute the concentrated solution. The driving force is the concentration gradient, and the magnitude of the driving force is measured as the osmotic pressure. It has been found that osmosis can be prevented by imposing a pressure gradient on the system. Furthermore, if the pressure gradient is larger than the osmotic pressure, water will actually be removed from the concentrated solution. This phenomenon is reverse osmosis (RO). The membrane consists of special polymers which allow water molecules to permeate the matrix but hold back other species. These membranes do not contain pores in the usual sense. They are usually made of cellulose acetate but can be made of nylon and other materials such as aromatic polyamides. The water is "soluble in the matrix," so it can pass through.

Very pure water can be obtained from reverse osmosis. The product water or permeate is almost pure water. Some other small species can permeate the membrane, but rejections of better than 90% can be obtained. In actual operation the permeate may be 75% of the input flow. Thus, a three to one reduction in the volume of effluent, a threefold increase in the concentration of suspended and dissolved solids, and a similar increase in concentration of the dispersed oily phase result. This decreases the interdrop distance and changes the ionic strength of the system, thus helping to destabilize the emulsion. At the same time the recovered water can be used for very critical applications. It makes an excellent boiler-feed and with a minimum of further treatment can be used for high-purity applications in the electronic industry.

In both electrodialysis and reverse osmosis the main fluid stream can pass through the system without having to pass through the membranes. In electrodialysis the ions and in reverse osmosis, the water, are withdrawn "at right angles" to the main flow. Thus, if care is taken in the design and in the operating parameters, the membranes do not become blinded. The straight-through flow can carry the suspended solids or liquid drops past the membranes and into the effluent.

Reverse osmosis systems can be of several types. The first system was a simple, plate and frame type which was followed by the spirally wound and hollow fiber configurations. Since the spacing of the spirally wound or hollow tube types is quite small, they are more prone to plug. The high surface area of the fiber types makes them excellent for the desalination of waters with little or no suspended matter, but the configurations with wider spacing are indicated for emulsions applications [2].

## ULTRAFILTRATION

Ultrafiltration (UF) membranes actually have very small pores. These membranes will pass water and some smaller dissolved materials of low molecular weight but will hold back large particles. Again, the main flow passes over the membrane at right angles to it, and the smaller species are removed. Because of the small pore size and the flow pattern, the membranes do not clog easily. In this process the emulsion is concentrated, and at the same time low-molecular-weight materials are removed. The recovered water is not as pure as in reverse osmosis, but it is free of oils and suspended matter and can often be reused. The concentrated emulsion can then be further treated. If a pH adjustment will destabilize it, less material will be required. If heating will help, there is less water to be heated. Smaller tanks are needed and other savings can be realized [3].

Hockenberry and Lieser give an interesting report on the advantages of ultrafiltration and reverse osmosis [4]. They state, "The standard means of disposal of oil wastewater has been chemical demulsification followed by secondary clarification of the water phase." By "chemical demulsification" they mean massive doses of sulfuric acid, alum, iron sulfate, or waste pickle acid. As mentioned before, these treatments result in large amounts of floc and sludge mixed with the oil which are difficult to resolve. They say, ". . . although these systems produce water phases suitable for discharge to sewers or public waterways, the oil recovered . . . is not readily recyclable. At best the oil can be used as a supplementary burner fuel."

These authors have run pilot tests on both reverse osmosis and ultrafiltration units [4]. Another interesting comment is, ". . . after a complete review of the literature . . . it was found that the treatability of a particular oily wastewater could not be predicted." They find that both reverse osmosis and ultrafiltration yield satisfactory oil which could probably be further treated and reused. The water obtained from the two processes is quite different, however. Both waters had less than 10 mg/l of mineral oil. The water from the ultrafiltration unit, however, had too much dissolved material to be released to the public waterway. It could, however, be treated in a conventional sewage plant to proper levels. They conclude that the lower cost of the UF units favors their use. If a real need exists for pure water or if the sewer charges are higher, the RO method might be preferable.

## FILTRATION COALESCENCE

If we increase the size of the pores in the membrane, we go from ultrafiltration to microfiltration and, finally, to actual filtration and fibrous-bed coalescers. Somewhere along the way we change from flowing the main stream past the filtration medium to passing it through the medium. This is conventional filtration. Here any material that is too big to pass the "holes" in the membrane is held on the matrix and must eventually be removed in some way. There are several ways that filtration can aid in demulsification. If the emulsion is stabilized by solids, it may be possible to filter out the solids and thus destabilize the emulsion. In practice this is done by passing the emulsion through a filter that is just fine enough to hold back most of the solids, but that is a coarse enough to allow a reasonably fast flow. This rate of flow will tend to wash off some of the oily material that might otherwise be mechanically held on the filter cake.

Finally, we come to the very coarse "filters" or coalescers. The action here is to force the droplets of oil into contact with each other or with an oil-wet surface and thus cause coalescence. Electrodialysis and reverse osmosis have been limited to aqueous, external phases. The membranes of ultrafiltration systems have also been designed for aqueous systems. Filters and coalescers, however, can be used on either oil-in-water or water-in-oil systems. "Hay tanks", which are a form of coalescer, have been used in oil fields for many years. The mode of action of coalescers has been the subject of many studies [5-10]. One of the first attempts to use filtration to resolve emulsions used two filter membranes as shown in Figure 6-2a. Here the emulsion is lead into a T-shaped device, one arm of which is closed by a water-wet filter and the other by an oil-wet filter unit. The system was designed to remove small amounts of water from an oil stream but could be set up to operate on water-external systems too. In either case the external phase passes freely through the filter medium that is wet by it. The suspended drops then rise (or fall) until they contact the other filter which they can pass through. In principle it sounds fine. In actual practice it is necessary to balance the system very carefully. If the rates of flow and the porosities of the filters are not carefully balanced, breakthrough can occur at either filter pack, and the system fails (Fig. 6.2b). Maintaining a proper balance is difficult, especially if there are suspended solids that cause the porosity of the system to change (Fig 6.2c).

Actual porous-bed coalescers do not try to hold back either phase. They act to force the suspended droplets into close contact and thus encourage coalescence. The coalescers are conceded to work best if the material of the filter unit is wet by the disperse phase. Then the drops are forced into contact with each other or

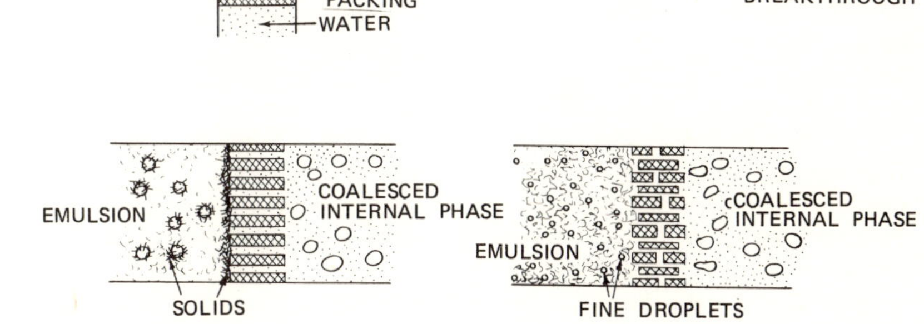


Figure 6.2 (a) T-cell filter separator. (b) Filter coalescence with breakthrough. (c) Coalescence by removal of solids. (d) Coarse filter coalescer.

the wet filter and form a compact second phase which moves through the filter and releases large drops on the downstream side. These large drops settle readily, and the coalesced phase can be removed from the system.

Coalescers can be of two general types depending on the type of packing. There are fibrous-bed types and those with particulate packing. Both types have been found to be effective in certain cases. With any type of filter coalescer the presence of filterable solids can plug the packing. The unit must then be repacked or backwashed. Many of the studies do not take this into account when assessing performance. Madea and his co-workers [5] find that an oil-wet packed bed is effective in coalescing oil-in-water dispersions. They report on tests of anthracite coal, Ottawa sand, powdered polypropylene, and a powdered polystyrene-type plastic. The test emulsion is a dispersion of Navy distillate (diesel-fuel type) oil in distilled water. The emulsion is passed through the packed column, and then any coalesced oil is allowed to separate. The ratio of the influent oil content to the settled effluent oil content is designated as the oil removal capability. Coal has a rating of about 1, sand and polyproylene have ratings of about 5 to 6, and the polystyrene type, a rating of about 25.

Spielman and Su [6] study the action of packed beds of glass beads. These glass beads are initially water wet. The authors postulate the presence of three differing regimes of the coalescing phase: first, oil microdrops suspended in a capillary-conducted, aqueous phase; then, held-up oil as discrete coalescing globules; and finally, capillary-conducted , oil-forming, well-connected channels. It is assumed that at the end of these channels the oil is released in large globules which can be easily separated and recovered (Fig. 6.2d). Merryman and Osterstock [7] report a series of tests on various Navy fuels. Their main purpose is to study the operation of parallel-plate coalescers. In these units the "filter pack" is a series of parallel plates. These plates can be of metal but, for the removal of oil, polypropylene is preferred because it is easily oil wet. The spacing on these plates is about 1 in. They are quite effective in coalescing free oil or loosely emulsified oil. If there is much detergent or other emulsifier present, they do not work well without a demulsifer chemical. The authors also find that, if the particle size is very small, a filter pack of polyurethane foam is needed to retard and coalesce the smaller particles.

Technical Report EHE-75-05 [8] covers a series of tests run on fibrous-bed coalescers. The media used were polyester felt, polypropylene felt, glass mats, and glass wool. A section filled with sand is also tested. The glass and the sand were tested in a water-wet condition and then treated with dimethyl silicone to render the oil-wettable and tested again. It is concluded that the oil-wet materials are more effective in removing small drops of oil from an aqueous stream. It is also noticed that it is necessary to build up a certain level of oil in the fiber pack before maximum efficiency is achieved. It is necessary to operate the unit at flow rates below a certain figure to minimize redispersion of the oil. It is concluded that the pertinent factors are filtering velocity, viscosity, temperature, porosity of the pack, and the fiber size.

Another paper studies the best way to separate the coalesced oil from the effluent stream [9]. It becomes apparent from the test that the optimum configuration is strongly dependent on the characteristics of the stream to be tested. It has repeatedly been found that if the interfacial tension is low the effectiveness of the coalescer may be poor. Some patents in this area are given in the bibliography at the end of the chapter [10-21].

## SLUDGE TREATMENT

It has been mentioned previously that many emulsion problems are not recognized as such. This is particularly true in the case of sludges. In many industrial situations a dilute water-in-oil emulsion or oil-in-water emulsion is allowed to settle or is centrifuged, and the internal phase is concentrated into a cream which will not coalesce and separate further. In other cases most of the disperse phase may separate leaving a layer of uncoalesced drops and debris at the junction of the two separated layers. These sludges, or interfaces, can be particularly troublesome.

Often the first stop in solving a sludge problem is to recognize that it really does involve demulsification. Then some simple tests can be performed to further establish the exact conditions.

A test procedure is needed that will resolve the sludge or interface into oil, water, and insoluble solids. One simple procedure is to mix a measured quantity of the material with an excess of a cosolvent such as acetone or ethanol. This procedure will often release the solids and either produce a single, liquid phase or two phases that can separate. While such a procedure is not practical on a commercial scale, it is very informative as a laboratory test. If the sludge is found to contain large amounts of solids, the next step is then to determine the source of the solids. If they are being produced by an upstream treatment with lime and alum or other inorganic flocculants, consideration should be given to using polymeric organic flocculants to reduce the solids load on the system. If the solids can be traced to a single source, it may be possible to isolate that substream and greatly reduce the sludge problem.

If the solids cannot be excluded from the system, it may be possible to arrange to selectively wet them so that they are carried in either the aqueous or oily phase and not concentrated at the interface.

It is often difficult to judge the nature of a sludge or interface from the superficial appearance. Concentrated emulsions can appear almost solid and still be composed of two fluid phases. Some sludges are multiple emulsions of oil-in-water-in-oil or water-in-oil-in-water. These sludges can look loose or lacy and still be very difficult to separate. One possible solution is to mix the sludge with additional external phase, heat it, and allow the material to settle while hot. The effectiveness of such a procedure can easily be tested on a laboratory scale. Once the solids have been isolated from the sludge, a commercial demulsifier or "slop oil compound" may be able to effect a separation.

Now before a final summation, let us take a close look at one particular field to see how their problems have been attacked and where the general principles we have been discussing apply. It should be remembered that those in the field do not regard their problems primarily as demulsification problems. To them it is a waste-disposal problem. On the other hand, perhaps looking at the field as a demulsification problem may let us see how to attack the concerns of other industries.

## SCOURING EFFLUENT

The area under consideration is the preparation of raw wool for textile use which is known as wool scouring. Raw wool can be obtained in two ways: by shearing sheep that are raised for their wool and by using the pelts of sheep that have been killed for meat. The wool that is obtained from the dehairing of the sheepskins, called pulled wool, is usually of poorer quality and often is used for other than woven textile purposes. We will concentrate on the processes whereby the sheared wool is scoured and readied for the spinning mills.

Raw wool actually contains only about 65% wool fiber. The rest is "yolk". The yolk can be divided into grease, suint, and dirt. Wool grease is a mixture of esters of branched chain and hydroxy acids plus the corresponding alcohols and cholesterol. When first deposited on the wool the material contains only a small amount of free fatty acid, but as it ages on the animal, it oxidizes and more free acids appear. Wool grease has a melting point of about 42°C.

If a portion of fleece is extracted with ether, the ether-soluble portion is called wool-wax. The water-soluble portion is called suint. Suint is believed to be the residue of the sweat of the sheep plus other water-soluble impurities that the fleece may trap. Suint is said to contain the potassium salts of fatty acids as well as sulfates and phosphates. Succinic, α-oxo-glutaric, glycolic, and levulinic acids have been found. The salts are said to be about 50-60% potassium, 4-5% sodium, and a small percentage of calcium [22]. The rest of the yolk is composed of insoluble solids particles such as fine sand, clay, and vegetable matter such as seeds and burrs.

We would like to separate the raw wool into clean fiber, the wool fats and greases, the water-soluble fatty acids, the potassium values, the inorganic solids, and the vegetable matter. We could then spin our wool fiber into cloth or make felt; the organic fatty esters and acids and alcohols could be used, as such, for synthesis or as pharmaceutical or cosmetic ingredients. The potassium could be used as a fertilizer ingredient or for other purposes. The sand and such could be used as land-fill with no ill effects, and the vegetable matter could be either composted or burned. Let us see how close we can come to this ideal, and what it might cost.

Primitive people cleaned the wool by rubbing it with clays (fuller's earth) and/or washing it in running water. Sometimes they used wood ashes or plant saponins as detergents. They then picked out the burrs by hand and carded the wool before spinning it.

As a wool-processing industry developed, the fundamental steps did not change too much. The raw wool is picked lightly to separate the fibers a little and then run through a series of vats (called bowls) where the wool is treated with aqueous solutions to remove the grease, solids, and suint. Three to seven bowls are used, and the water moves from bowl to bowl countercurrent to the flow of the wool. Between each step the wool passes through rollers that press out much of the water. The first bowl originally contained a warm soap solution. Then it was found that, if soda ash was added, it would form soaps from the fatty acids of the wool grease. As the process became more sophisticated, pH control was instituted, and then synthetic detergents were introduced. Thus, clean water is used in the last rinse of the wool. This rinse water is used in the next to last rinse, and the water moves up until very dirty water is discharged from the first bowl. This effluent contains emulsified wool-grease, suint, suspended dirt, and detergent.

It was once the practice to "acid-crack" this waste water by adjusting the pH to about 3 or 4 with sulfuric acid and then boiling the whole thing. This treatment partially broke the emulsion, and the mixture was then centrifuged. The solids came out as a heavy end, the oil came out as a cream, and a relatively clean, middle fraction was then dumped. Some lanolin or wool fat was recovered from the cream. The scoured wool was then treated with more sulfuric acid and heated hot enough to cahr any vegetable material. The wool was then pressed to break up the char and rinsed. The acidulated wash-waters were sometimes neutralized with lime before dumping. As the loads on the rivers became heavier and their natural recovery rates were exceeded, pressure developed to stop this dumping of untreated waste. Since any treatment meant more costs, it was the practice to do just what was required by law. The waste was neutralized and dumped to the municipal sewers. When these sewers became overloaded, charges were put on dumping. Thus, a gradual, patchwork approach evolved which just kept ahead of the complaints.

At the same time the technical people in the wool-treating industry recognized that something really effective had to be done. Several detailed and careful programs were initiated to try to solve the problem. The situation does not seem to have been recognized as a demulsification problem but rather treated as a solids removal problem. The first studies concentrated on cleaning up the effluent water and getting most of the solids and grease out as a sludge. The water was then partially reused or biologically treated to meet disposal requirements. The sludge was the big problem. Acid-cracking would recover about 30-40% of the wool grease which could be solvent-refined or sold. The rest of the sludge was hauled out to sea or spread on land as fertilizer. Incineration was also tried. At Bradford, England, a huge, sewage-treatment plant was built to treat the effluent from a group of mills as well as domestic sewage. The present-day, total, dry-weather flow is in excess of 20 million gal/day. This is all acid cracked. The sludge produced is dewatered in 128 sludge presses, yielding over 800 long tons of press cake a week. The press cake was first used as a soil conditioner but is now being incinerated [23].

Let us now apply some of the principles covered in earlier chapters to this problem and see if we can get any guidance. Such a study will also serve to illustrate how to attack a similar problem. This case involves an emulsion of oil-in-water. The emulsion contains about 5% oil (wool grease) and about 3% suspended solids while the external, aqueous phase contains about 4% dissolved solids. The dissolved solids are potassium salts of organic acids along with urea and other fecal solubles. The emulsifying agents are sodium and potassium soaps and the detergents used to wash the wool. Perhaps we can help the situation by using a different detergent. This approach has been tried. Anionic and cationic detergents have been used and now nonionics are in favor. They are effective in cleaning the wool but do not greatly help the clean-up problem. Perhaps we don't need nay additional detergent at all. There are free organic acids in the wool grease. If we add alkali, we will make soaps. This is now being done to a certain extent. It does save money and can be made to yield clean wool. It doesn't solve the waste problems.

We know that if we change the ionic strength of the system, we may destabilize the emulsion. This procedure has been tried. The South African Wool and Textile Research Institute (SAWTRI) has done very detailed studies in this area. Work reported by Trupie and Mozes started with the use of added seawater to destabilize the system. They found that this can be of real help [24,25]. Mozes then set out to determine what constituents of seawater were important [26]. It was found that sodium chloride, magnesium chloride, and magnesium sulfate were the important salts. Pilot-plant scale work confirmed the laboratory conclusions [27,28]. The next step was to see if other similar materials would have the same effect. The magnesium-rich waste liquor from a salt (NaCl) recovery plant (bitterns) was tested. This material worked even better [29]. Again pilot-plant tests confirmed the results [30]. For those plants without a source of bitterns it was found that magnesium chloride could also be used [31].

Meanwhile, the same authors had made a careful study of the particle size distributions of the solid dirt particles [32]. They found that a 3-μm filter would reject 95% of the particles. They also studied the flocculation of the effluent using alum and polyelectrolytes [33,34]. This approach gave poorer results and required careful pH control [35]. However, it did remove dirt well [36].

It has been mentioned that dilute emulsions are often harder to break than more concentrated ones. Perhaps we should try to concentrate the scour liquor. One way to do this would be by ultrafiltration. This would separate the flow into two parts. There would be a main stream that contained the grease and solids and a side stream that contained most of the water and water-soluble components. This, too, has been tried [36-39]. Pearson, Anderson, and Wood showed that ultrafiltration of wool-scouring liquor could yield a clear solution with about 80-85% of the original volume and a sludge containing all the suspended solids and grease [38]. The clear liquid contained almost all the suint. They assumed that it would be treated further in a conventional sewage system. One is lead to ask if reverse osmosis would allow the reuse of the water and would concentrate the suint to the point where the potassium values or the organics could be recovered.

Tests have also been run on the use of membranes with larger pores (microfiltration) [40,41]. These systems showed somewhat faster separations, but more grease was passed through the membrane. Solids were removed with a rejection factor of 90% or better, but in some cases the grease rejection factor was as low as 77%.

Another way to break an emulsion is to add to the system a solvent that will destabilize the interface. This approach has been attempted with some success [42]. McLachlan, Smith, and Fields report that the combination of a synthetic polyelectrolyte and n-pentanol results in rapid separation of a layer containing the grease dirt, and alcohol along with most of the detergent. If this layer is heated to 80-90°C for 30 min, it separates into three layers. The top layer contains the grease, excess alcohol, and detergent. The middle layer contains flocculated dirt. The bottom layer contains an aqueous phase saturated with n-pentanol. The n-pentanol was recovered by steam stripping.

So it can be seen that most of the methods for resolving the wool-grease emulsion have been tried. What actually is the current practice in the mills? Here again it is necessary to realize that the object of the mill is to get as much good, clean wool as possible for the least expenditure. From the standpoint of economics, the only reason to do anything with the effluent is either to recover some of the grease or other materials at a profit or to avoid large sewage charges or fines. Thus, the cheapest way to solve the problem is to spread the effluent on the ground as a soil conditioner and fertilizer. It has been shown that this can be done, but the amount of liquor that can be utilized by an acre of ground is limited, and the odor from the process is considerable. Thus, the use of the system is limited to rural areas with large tracts of land.

## WRONZ SYSTEM

Gibson, Morgan, and Robinson, in a 1979 article [43], outline the present status of the problem very well. They point out that the rising cost of clean water and the effluent-treatment charges assessed have forced the wool-scouring industry to look at better ways to operate their plants. Two general systems are now being adopted—the (Wool Research Organization of New Zealand) WRONZ system and the Commonwealth Scientific and Industrial Research Organization Lo-flow (CSIRO) process.

In the WRONZ system the effluent is drawn from the first two bowls at timed intervals, and the heavy solids are separated in a paralleled-inclined-plate separator [43]. The stream is then heated and degreased in a disk centrifuge. The flow is then cycled back to the first bowl. In this system the emulsion is destabilized by allowing the concentration of suint to build up in the first two bowls. The grease from the centrifuge is then often run to a GARAP grease-recovery centrifuge.* This system is said to be a great improvement on previous methods. It still results in recovery of only about half of the grease. Effluent from this system is then further treated in a conventional biological system.

The WRONZ system has resulted in greatly reduced water consumption and lower effluent volumes [43]. Also less grease, less heavy solids, and less fiber are present in the waste. The effluent is still, however, an emulsion of grease in water and must be further treated. Four approaches are being studied. Here again there is a reversion to solids-treatment methods. The chemical flocculation that they try consists of treatment with hot alum or hot acid-cracking. As usual, they find that treatment with alum results in a large floc volume. The hot acid-cracking system is then studied. Batch treatment of effluent has been used for a long time, but the odor of the open boiling vats and the storage ponds or tanks has been the object of much attack. A continuous, closed system has been devised and tested. It uses polyelectrolyte flocculation. The grease and solids are sent to a landfill.

Ultrafiltration was tested and found to give an acceptable effluent, but the concentrate was not in a form that can be easily handled. [43]. Apparently no consideration is being given to recovery of the grease values.

Evaporation and incineration of the sludge were also tested [43]. The authors says, "This process [evaporation and incineration] is the ultimate solution for the treatment of wool scour effluent."

One wonders how long it will be before raw material shortages will make full recovery of waste materials not only economic but also essential. In any event, pilot tests show that a steam-evaporated concentrate could be incinerated with a minimum of auxillary fuel [43].

*Marketed by A. Thibeau et Cie of France.

An interesting biological system is also mentioned [43]. In this system a shaft is drilled into an impermeable formation and lined with a suitable material. Sludge is then pumped into the shaft at the bottom, and an anaerobic biological decomposition is set up, taking advantage of the hydrostatic pressure. The sludge from this process can be sterilized and used as a fertilizer.

## LO-FLO SYSTEM

The Lo-flo system is based on the use of concentrated suint to destabilize the emulsion. The Lo-flo unit replaces the first bowl in a conventional scouring line. It consists of three small bowls, two of which are jet-scour bowls rather than immersion bowls. The liquor in the unit is recirculated until the level of dissolved solids destabilizes the emulsion, then the insoluble solids are removed in a decanter certifuge. The sludge from this step has a relatively low biological oxygen demand (BOD) and is at least 50% solids. The grease is then removed in a special, grease-separator centrifuge. Over 90% grease recovery is claimed. The liquor is then returned to the Lo-flo unit. Finally, an equilibrium is reached where the solids and grease are removed, and a stable scouring-liquor attained. The only effluent is that carried by the solids or grease and that carried on the wool to the second conventional bowl. The effluent from this bowl has very little BOD or grease and can be handled by a normal sewage treatment plant. The volume of effluent is greatly reduced and, indeed, it is claimed that the introduction of a second Lo-flo unit in place of the second bowl achieves total recycle [44]. The grease recovered from this process has very little additional material in it. It can be solvent-treated and sold at a profit.

Each of these processes has certain costs associated with it. Some require little capital investment but have high operating costs. Others show good operating economics but require high initial capital investment [45]. Gibson, Morgan, and Robinson report that the Lo-flo system seems to have the lowest overall costs [43].

## SUMMARY

Now that we have gone through an effluent problem in considerable detail and shown how the basic principles have been applied and how the concepts of demulsification have to be incorporated into a total system, it should be clear that these principles can be applied to any area.

## REFERENCES


1. R. P. Ouellette, J. A. King, and P. N. Cheremisinoff, *Electrotechnology*, Vol. 1, Ann Arbor Science Publishers Inc., Ann Arbor, Michigan, 1978 pp. 15-39
2. R. P. Ouellette, J. A. King, and P. N. Cheremisinoff, *Electrotechnology*, Vol. 1, Ann Arbor Science Publishers Inc., Ann Arbor, Michigan, 1978 pp. 41-74
3. R. P. Ouellette, J. A. King, and P. N. Cheremisinoff, *Electrotechnology*, Vol. 1, Ann Arbor Science Publishers Inc., Ann Arbor, Michigan, 1978, pp. 75-99
4. H. R. Hockenberry and J. E. Lieser, *J. Soc. Lub. Eng., 33*: 247-251 (1977)
5. J. R. Madea, S. M. Fruh, C. A. Miller, and A. Beerbower, *Environ. Sci. Tech. 10*:1044-1046 (1976)
6. L. A. Spielman and P. S. Yeang, *Inc. Fng. Chem. Fundam. 16*:272-282 (1977)
7. J. S. Merryman and E. R. Osterstock, AD758319, NTIS (1973)
8. J. N. Chieu, R. S. Schecter, M. J. Humenick, and E. F. Glayna, *Tech. Report CRWR-136, EHE75-05,* U. of Texas, Austin, TX (1975)
9. W. H. Langdon, P. P. Naik, and D. T. Wasan, *Envirn. Sci. Tech. 6*:905-910 (1971)
10. C. C. J. Wu, R. S. Schechter, J. F. Malina, and E. F. Glayna, *Tech. Report CRWR-143, EHE77-01,* U. of Texas, Austin, TX (1977)
11. T. Amano and M. Shendo, Jap. Patent 74-114,268, CA *84*: 217548
12. Y. Yamase, M. Ikeguchi and Y. Nagata, Jap. Patent 75-96,481 CA *84*:21846e (1975)
13. K. Kawachi, Y. Negi and S. Nositani, Jap. Patent 7,514, 576 CA *84*:8768m (1975)
14. A. Kobayashi, M. Kumazawa and K. Tsuda, Jap. Patent 75-74,571 CA *84*:8768m (1975)
15. S. Kita, T. Kono and K. Kawai, Jap. Patent 75-40, 481, CA *84*:21775ff (1975)
16. T. Sakamato and T. Amemiya, Jap. Patent 75-01087, CA *84*: 46609t (1975)
17. D. W. Beavon, Brit. Patent 1,340,931 (1973)
18. A. Gallan and D. H. Fruman, Ger Patent 2,354,939 CA *83*: 47918j (1974)
19. L. W. Jones, U.S. Patent 3,893,925 (1975)
20. H. F. Keller, Jr., U.S. Patent 3,867,285 (1975)
21. R. H. Terhune and J. C. Duke, U.S. Patent 3,865,732 (1975)

22. A. H. Mahsin and S. M. Shah, *Pak. J. Sci. Ind. Res. 12*: 286-293 (1969)
23. B. M. Rack, *I. Mechn. E. Conf. Proc.* 1973-3:77-78 (1977)
24. B. M. Rack, *I. Mechn. E. Conf. Proc.* Report 339 (1977)
25. D. W. F. Turpie and T. E. Mozes, *SAWTRI Tech. Rep. #343, Treatment of Wool Scouring Liquors Part IV* (1977)
26. T. E. Mozes, *SAWTRI Tech. Rep. #344 Part V* (1977)
27. D. W. F. Turpie, T. E. Mozes, *SAWTRI Tech. Rep. #363*
28. T. E. Mozes, D. W. F. Turpie, *ASWTRI Tech. Rep. #411*, Part X (1978) SAWTRI, Port Elizabeth, So. Af.
29. T. W. Moxes, SAWTRI Tech. Rep. #422, *Treatment of Wool Scouring Liquors Part XI*, (1978)
30. T. E. Mozes, N. J. S. van Rensburg, D. W. F. Turpie, SAWTRI Tech. Rep. #467, *Treatment of Wool Scouring Liquors Part XIII* (1981)
31. T. W. Mozes, N. J. S. van Rensburg, D. W. F. Turpie, SAWTRI *Tech. Rep. #468, Part XIV* (1981)
32. D. W. F. Turpie and T. E. Mozes *SAWTRI Tech. Rep. #384* Part VIII (1977)
33. T. E. Mozes, *SAWTRI Tech. Rep. #338 Treatment of Wool Scouring Liquors Part 1.* (1977)
34. T. E. Mozes, D. W. F. Turpie, *SAWTRI Tech, Rep. #338* loc at #11 (1977)
35. J. R. Cristoe, *J. Water Pollution Control Fed. 49*:848-854 (1977)
36. C. N. S. McLachlan, D. K. W. Smith and R. J. Webb. *Water Research 14*:729-733 (1980)
37. T. E. Mozes and D. W. F. Turpie, *SAWTRI Tech. Rep. #349*
38. A. J. C. Pearson, C. A. Anderson and G. F. Wood, *J. Water Pollution Control Fed., 48*:945-953 (1976)
39. P. E. Chisnall and B. Robinson, *J. Text. Inst. 9*:397-406 (1979)
40. T. E. Mozes and D. W. F. Turpie, *SAWTRI Tech. Rep. #386* Part IX (1978)
41. T. E. Mozes and D. W. F. Turpie, *SAWTRI Tech. Rep. #453,* Part XII (1979)
42. C. N. S. McLachlan, D. W. K. Smith and R. B. Fields, *Water Research 14*:735-740 (1980)
43. J. M. D. Gibson, W. B. Morgan and B. Robinson, *Textile Institute and Ind. 17*:31-37 (1979)
44. G. F. Wood, A. J. C. Pearson, J. R. Cristoe, *CSIRO Report G39,* ISBN#0-643-02465-4
45. B. R. Evans, *Effluent and Water Treat. J., 14*:85-88 (1974)

# Author Index

*Numbers in brackets are reference numbers and indicate that an author's work is referred to although his name is not cited in the text. Italic numbers give the page on which the complete reference is listed.*


Adamson, A. W., 25[32], *36*
Alden, S. F., 117[27], *133*
Allen, W. B., 128[65], *134*
Allenson, J. T., 81[44], *90*
Amano, T., 143[11], *151*
Amemiya, T., 143[16], *151*
Anacker, E. W., 23[24], *36*
Anderson, C. A., 148[38], *152*
Anon, 81[50], *91*, *102*[4], *103*
Aronson, M. P., 32[56], *37*, 83[60], *91*
Arpesella, W., 64[10], *89*
Asafagan, I., 81[53], *91*
Aughey, W. H., 44[28], *61*

Balden, A. R., 66[20], *90*
Bambah, R. P., 31[46], *37*
Bancroft, W. D., 4[1], *35*
Bansbach, P. L., 112[8], *133*
Barcus, W. J., 55[43], *62*
Barden, R. E., [72], *38*
Barker, J. E., 79[32], *90*
Barnickel, W. S., 120[61,63], *134*
Baum, F. J., 44[28], *61*
Beaucaire, V. D., 79[31], *90*
Beavon, D. W., 143[17], *151*
Becher, P., 44[16], *61*
Beerbower, A., 141[5], *151*
Bellocq, A. M., [67], *38*
Biaes, J., [67], *38*
Blair, C. M., 112[10], *133*
Boardman, R. P., [71], *38*
Bolsterle, M., 31[49], *37*
Borys, N. F., [72], *38*
Bradley, G. M., 27[36], *36*
Brady, G., 21[14], *36*
Brady, J. D., 117[24], *133*
Bridie, A. L., 131[69], *134*
Brisford, R. H., 30[39], *37*
Bron, J., 19[11], *35*

Cage, J. M., 118[41], *134*
Campbell, A. N., 25[28], *36*
Canevoir, G. P., 132[70], *134*
Carmel, E. J., 129[67], *134*
Castornia, S., 66[19], *90*
Chantereau, R., 64[4], *89*
Cheever, A., 116[21], *133*
Cheremisinoff, P. N., 139[3], *151*
Chieu, J. N., 81[41], *90*, 141, 142[8], *151*
Chisnall, P. E., 147[39], *152*

Clift, R., 47[36], *61*
Olin, D., [07], *38*
Coexter, J. S. M., 32[52], 31[45], *37*
Cooper, D. G., 129[67], *134*
Cottrell, A. G., 116[23], *133*
Cottrell, F. G., 95[1], *103*, 115[20], *133*
Coulter, J. B., 65[18], *90*
Crandall, P. G., 64[11], *89*
Crist, W., 81[52], *91*
Cristoe, J. R., 147[35], *152*
Crompton, F. M., 65[18], *90*
Cyprian, J., 81[52], *91*

Dailey, J. H., 64[3], *89*
Davidson, D. D., 84[40], *90*
Davis, J. T., [68], *38*
Debye, P., 23[23], *36*
Degroote, M., 120[64], *134*
Dodd, A. V., 120[60], *134*
Dodds, J. A., 31[43], *37*
Douglas, P. L., 66[22], 79[26], *90*
Dow, D. B., 112[4], 113[11], *133*
Dressel, E. G., 65[13], *89*
Duke, J. C., 143[21], *151*
Dzieciuch, M. A., 81[48], *91*

Eddy, A. C., 115[19], 117[27, 31], *133*, 118[38], *134*
Eddy, W. G., 115[19], *133*
Eicke, J. F., 27[33], *36*
Einstein, A., 11[9], *35*
Elder, M. E., 44[25], *61*
Elworthy, P. H., 128[66], *134*
Endter, F., 44[20], *61*
Espenhorst, H., 64[2], *89*
Evans, B. R., 150[45], *152*

Farr, H. V., 66[21], *90*
Few, L., 31[45], *37*
Fichtmueller, R., 79[29], *90*
Fields, R. B., 148[42], *152*
Findlay, A., 25[27], *36*
Fisher, H. F., 114[12], 118[36], *133*
Florence, A. T., 128[66], *134*
Flumerfelt, R. W., [61], *37*
Fodiman, N. M., 65[16], *89*
Fogler, H. S., 54[41], *62*
Fruh, S. M., 141[5], *151*
Fruman, D. H., 143[18], *151*

Gallan, A., 143[18], *151*
Galloway, T. R., 65[12], *89*
Gardner, M., 30[42], *37*
Gaser, A., 81[51,52], *91*
Gealer, R. L., 81[48], *91*
Gebauer, H., 44[20], *61*
Gelat, A., [67], *38*
Gibbs, J. W., 25[30], *36*
Gibson, J. M. D., 148, 150[43], *152*
Gindler, E. M., 44[19], *61*
Glasstone, S., 31[31], *36*
Glayna, E. F., 141[8,10], 143[10], *151*, 81[41], *90*
Goodarz-Nia, I., 11, 14[10], *35*, [62], *37*
Grace, J. R., 47[36], *61*
Grave, G., 118[53], *134*
Graves, R. W., [57], *37*
Greenlee, R. W., 115[18], *133*, 119[58], *134*
Gummels, H., 65[17], *90*
Gupner, O., 118[51], *134*

Hall, J., 40[4], *60*
Harrel, J. W., 128[65]. *134*
Harrington, C. D., 66[21], *90*
Harris, F. W., 117[28], *133*
Hartland, S., [59], *37*
Hartman, R. J., 40[3], *60*
Harusawa, F., 27[34], *36*
Hayaski, S., [70], *38*

Hazeltine, T. R., 65[14], *89*
Hellis, W., 44[22], *61*
Hemming, M. L., 81[46], *91*
Henley, T. D., 79[30], *90*
Herbsman, A. M., 118[40], *134*
Hess, K., 23[17], *36*
Higuchi, W. I., 46[31], *61*
Hoar, T. P., 23[18], *36*
Hockenberry, H. R., 139[4], *151*
Hoffman, W. H., 80[38], *90*
Holt, S. L., [72], *38*
Humenick, M. J., 141[8], *151*

Ikeda, S., [70], *38*
Ikeguchi, M., 143[12], *151*
Imae, T., [70], *38*
Isemura, T., 23[20], *36*
Ishkahanova, E. B., 64[1], *89*
Ismail, M. A., 64[11], *89*

Jackson, D. E., 115[15], *133*
Jarvis, A. R., 118[44], *134*
Jones, L. W., 143[19], *151*

Kawachi, K., 143[13], *151*
Kawai, K., 141[15], *151*
Kaye, B. H., [71], *38*
Keller, H. F., 143[20], *151*
Kerker, M., 44[23], *61*
Kertes, A. S., [69], *38*
Khroniykh, V. F., 64[9], *89*
King, A., 42[17], *61*
King, J. A., 139[2], *151*
Kinney, C. R., 65[15], *89*
Kirschenbauer, H. A., 95[2], *103*
Kita, S., 143[15], *151*
Kobayashi, A., 143[14], *151*
Kono, T., 141[15], *151*
Koretski, A. F., 27[35], *36*
Krishnan, R., [65], *37*
Kucher, R. V., 23[25], *36*
Kumazawa, M., 143[14], *151*
Kuriyama, K. J., 23[26],
Kurland, R. S., 81[42], *90*
Kyger, J. A., 66[21],

Lai, W. C., [69], *38*
Laird, R. E., 117[29], *133*
Lalanne, P., [67], *38*
Lamer, V. K., 46[33], *61*
Langdon, W. H., 141, 143[9], *151*
Langese, J. F., 64[18], *90*
Lauermann, G., 80[36], *90*
Lawrason, L., 118[39], *134*
Lawrence, A. S. C., 98, 99[3], *103*
Lebowitz, J. L., 31[48], *37*
Lein, T. R., 46[35], *61*
Leiser, J. E., 140[4], *151*
Lemanceau, B., [67], *38*
Lissant, K. J., 8[3], *35*, 23[15], 27[37], *36*, 30[40,41], 32[54], *37*, 42[15a], *61*, 83[54,55,58], *91*
Livamov, V. A., 64[9], *89*
Loison, R., 81[49], *91*
Lucas, R. N., 115[17], *133*, 119[59], *134*
Luthy, R. G., 65[12], *89*

Macley, W. N., 44[19], *61*
Madea, J. R., 141, 142[5], *151*
Maeda, H., [73], *38*
Mahsin, A. H., 145[22], *152*
Malina, J. F., 141, 143[10], *151*
Mandelcorn, L., 21[14], *36*
Maron, S. H., 44[24], *61*
Maser, J. C., 32[56], *37*
Mason, G., 31[47], *37*
Mathews, P., 64[5], *89*
Matuera, R., [75], *38*
Matzke, E. B., 32[51], *37*

Mavlyutova, M. Z., 81[53], 91
Mayhan, K. G., 32[53], 37, 83[56], 91
Mc Bain, M. E., 23[16], 36
McBain, J. W., 23[21,22], 36
McKibben, C. W., 188[34] 133
McLachlan, C. N. S., 147[36], 152
Megal, L. D., 81[47], 91
Meinzer, G. H., 114[13], 133
Meredith, W., 117[26], 133
Merryman, J. S., 141,142[7], 151
Midlam, R. R., 66[21], 90
Mie, G., 44[26], 61
Militsin, S. V., 79[28], 90
Miller, C. A., 141[5], 151
Mitsui, T., 27[34], 36
Mittal, K. L., [58], 37
Mongait, I. L., 64[1], 89
Morgan, W. B., 149[43], 152
Moroi, Y., [75], 38
Morrell, J. C., 118[37], 133
Mozes, T. E., 147[26,27,29, 37], 148[41], 152
Mukherjee, L. N., 44[17], 61
Muller, K. W., 9[7], 11[7], 35
Musner, J., 80[36], 90
Mysjokin, B. A., 64[9], 89

Nagaia, Y., 143[12], 151
Naggi, F., 46[10], 89
Naik, P. P., 141[9], 151
Nakagawa, T., 23[20], 36
Nakai, K., 81[45], 91
Nakamura, S., 27[34], 36
Negi, Y., 143[13], 151
Nemethy, G., [74], 38
Newman, J., 40[1], 60
Ninhan, B. W., [66], 38
Nositani, S., 143[13], 151

Ober, R., [63], 37
Ockert, K. F., 65[15], 89
Okazawa, T., 19[11], 35, 19[11], 20[11], 35
Orelup, 40[6], 60
Oshima, T., 64[7], 89
Osterstock, E. R., 141[7], 151
Ouellette, R. P., 137[1], 151

Panich, R. M., 65[16], 89
Peace, B. W., 32[54], 37, 83[57], 91
Pearson, A. J. C., 147[38], 152
Phillips, W. J., [57], 37
Phillips, C. R., 46[35], 61
Pilat, J., 64[8], 89
Pincus, I., 65[15], 89
Princen, H. M., 32[55], 37 83[59,60] 91

Quincke, G., 4[2], 35

Rack, B. M., 146[23], 147[24], 152
Raney, J. H., 117[30], 133
Reddy, S. R., 54[41], 62
Reiss, H., 20[13], 35
Ricci, J. E., 29[29], 36
Riley, P. P., 23[19], 36
Robertson, T. B., 40[8], 60
Robinson, B., 147[39], 152
Rogers, C. A., 31[44,45], 37
Rogers, J. A., 128[66], 134
Rosenbaun, W. C., 66[21], 90
Rowe, E. L., 46[32], 61
Roy, A., [74], 38
Ruckenstein, E., 19[12], 35, [64], 37
Rurness, C. D., 81[44], 90
Rusz, A., 11[8], 35

Rybachok, I. N., 118[48], *134*

Sakamato, T., 143[16], *151*
Samocha, T., 79[27], *90*
Schechter, R. S., 81[41], *90*, 141[8,10], 143[10], *151*
Schilinski, H., 79[29], *90*
Schrnezel, M., 46[34], *61*
Schulaman, J. H., 23[18], *36*
Schulz, J., 80[36], *90*
Scott, J. C., 115[15], *133*
Sedlmeiler, H., 81[43], *90*
Seibert, F. M., 117[24], *133*
Selleck, R. E., 65[12], *89*
Shah, S. M., 145[22], *152*
Shea, G. B., 112[5], *133*
Shendo, M., 143[11], *151*
Sherman, P., 40[7], *60*
Shinoda, K., 23[20], *36*
Shirly, E., L., 118[50], *134*
Sibree, J. O., 44[18], *61*
Skul ´Sku, V. M., 79[28], *90*
Sloan, C. K., 44[27], *61*
Smalley, I., 31[50], *37*
Smirnova, A. V., 27[35], *36*
Smith, D. K. W., 147[36], *152*
Smith, N. O., 25[28], *36*
Sokalovskaya, N. A., 27[35], *36*
Sole, P., 80[37], *90*
Somer, T. G., 80[34], *90*
Speed, J. B., 116[22], *133*
Spielman, L. A., 141, 142[6], *151*, 142[6], *151*
Spina, N., 64[10], *89*
Stausser, J. R., 81[42], *90*
Steinhauff, R., 122[7], *133*
Stevenson, A. F., 44[22], *61*
Stigter, D., [60], *37*
Stock, H., 118[52], *134*
Strassner, J. F., 129,130, 131[68], *134*

Takayanagi, S., 64[7], *89*
Takeneshi, S., 80[39], *90*
Tamamuski, B., 23[20], *36*
Taupin, C., [63], *37*
Terhune, R. H., 143[21], *151*
Thayer, H. E., 66[21], *90*
Thoenes, H. W., 81[43], *90*
Thompson, R. S., 79[32], *90*
Toth, R., 30[38], *37*
Tscherner, J., 79[29], *90*
Tsuda, K., 143[14], *151*
Tsunoda, M. A., [73], *38*
Turner, D. W., 118[42], *134*
Turner, D. N., 118[49], *134*
Turpie, D. W. F., 147[25,28], 148[40], *152*

Uran, L. G., 112[6], *133*
Vabra, D. K., [59], *37*
Van Der Heiyde, H. B., 131[69], *134*
Van Rensburg, N. J. S. 147[30], *152*
Visscher, W. M., 31[49], *37*
Vold, R. D., [58], *37*
Voltz, V. W., 79[32], *90*
Von Smoluchowski, M., 9[4,5], *35*
Voyulskij, S. S., 65[16], *89*

Wachtel, R. E., 46[33], *61*
Wales, M., 44[21], *61*
Wallach, M. L., 44[22], *61*
Wanders, T. H., 131[69], *134*
Wasan, D. T., 141[9], *151*
Waterman, L. C., 115[16], *133*, 118[50], *134*
Watson, F. D., 118[43,45], *134*
Webb, R. J., 147[36], *152*
Weber, M. E., 47[36], *61*

Webster, W. W., 128[65], *134*
Woidmann, H., 80[33], *90*
Weigner, G., 9[6], *35*
Weintraub, M. H., 81[48], *91*
Welch, S., 54[40], *62*
Wibbles, H. L., 66[21], *90*
Wiggins, J. L., 112[9], *133*
Winkler, C. A., 21[14], *36*
Winslow, S. D., 115[16], *133*
Woelfin, W., 118[47], *134*
Wood, G. F. 147[38], 150[44], *152*
Wood, J. W. 129[67], *134*
Woodman, J., 40[2], *60*
Worthington, J. T., 117[25], *133*
Wristow, J. D., 118[43], *134*
Wu, C. C. J., 141, 143[10], *151*
Wyatt, G. H., 64[6], *89*

Yamase, Y., 143[12], *151*
Yeager, H., 66[21], *90*
Yeang, P. S., 141[6], *151*
Yurzhenko, A. I., 23[25], *36*

Zakahrina, S. B., 64[1], *89*
Zajic, J. E., 129[67], *134*
Zegveld, W., 131[69], *134*
Zomick, D., 31[48], *37*

# Subject Index

Additional Methods and Areas in Demulsification, 135
Airplane Wash-Waters, 65
API Separator, 74
Appearance, 44
Asphaltenes, 131

Babcock Test, 42
Backflow, 34
Bench Scale Dissolved Air Flotation Unit, 56
Bench Scale Flotation Tests, 56
Bilge Water, 65
Bottle Tests, 121
Bottom Settings and Water (BS&W), 42
Break an Emulsion, 30
Breaking Methods, 94
Brilliant Blue FCF, 41
Brownian Movement, 11, 13, 14, 15, 16
BS & W Testing, 108

Centrifuge, 30, 42, 113
Chemical Potential, 20
Chemical Demulsification, 99-119
Chemicals Used in Demulsification, 87
Chocolate Mousses, 131
Circular Separator, 74, 76
Citrus oils, 79
Citrus-Processing Plants, 64
Close Packing of Drops, 85
Closest Packing of Spheres, 31
Cloudy Oils, 93
Coagulation Rate, 11
Coalescence, 5, 16, 34
Coalescers, Fibrous Bed, 142
Collision Effectiveness, 9, 11
Components and Amounts of Each Phase, 41
Composition Spaces, 26
Computer Simulation, 17
Conductivity Testing, 41
Constructs, 4
Cosurfactants, 23
Coulter Counter, 45, 46
Creaming, 30, 53
Critical Micelle Concentration, 23
Crude-Oil Reservoir, 105
Cube, 31
Cyclone Separators, 71, 72

Cylindrical Model, 23, 32

Deane-Starke Method, 42
Degrees of Freedom, 25
Demulsification of Oil-in Water Emulsions, 63
Demulsification of Petroleum Emulsions, 105
Demulsification of Water-in-Oil Emulsions, 93
Desalting, 119
Design Considerations, 76
Diffusion Coefficient, 10, 11
Dispersion Mechanisms, 47, 49
Dissolved Air Flotation (DAF), 55
Dodecahedra, 32
Double Bee Cell, 32
Drilling-Oil Emulsions, 65
Drop Spacing, 6, 7
Drop Breakup Mechanisms, 50
Dye Test, 41

Einstein's Equation, 11, 14
Electrical Dehydration, 115
Electrical Demulsification, 114
Electrical Fields, 52
Electric Field, 138
Electrodialysis, 136
Emulsifying Agents, 21
Emulsion, Definition, 4
Emulsion Types, 40, 135
Emulsions, Spectroscopy, 42
Evaporation, 149
External Phase, 39, 40, 46

Fast Mix Cycle, 54
Filters, 113
Filtration Coalescence, 136, 140
Fire-Floods, 126
Flocculation, 5, 30, 53, 78
Flocculation Chamber, 57
Flocs, 33
Flotation Methods, 53
Foams, 32
Freon, 113, 54
Full-Pressurization Test, 57

Gas Separators, 109
Geometry, 30
Gibbs Free Energy, 20
Grease, 144
Grinding Waters, 64
Gun Barrels, 110

Heater Treaters, 111
Heating, 42
Helmholtz Free Energy, 20
High-Internal-Phase Emulsions, 32
HLB Values, 129
Huff and Puff System, 126

Immiscible Liquid, 25
Incineration, 149
Indices of Refraction, 43
Induced Air flotation (IAF), 55, 57
Induced Air Flotation Unit WEMCO, 58
Induced Air Flotation Unit, 59
Injection Pump, 123
Interface, 2, 35, 93
Interface, Liquid-Liquid, 2
Interfacial Area, 20
Interfacial Phase, 23
Interfacial Film, 29
Interfacial Tension, 3, 20, 129
Internal Phase, 39
Inversion, 32, 83
Inversion and Phase Reversal, 83
Invert Emulsion, 4
Ion Exchange

Jar Test, 53, 54, 55, 78
Lamellar Micelles, 23
Lease Automatic Custody Transfer (LACT), 123
Light-Scattering Data, 44
Liquids, 2
Lo-Flow (CSIRO) Process, 149, 150
Low-Gravity Crudes, 125

Mechanical Coalescence, 81
Mechanical Work, 20
Membrane, 138
Membrane Techniques, 102
Micellar Region, 29
Micellar Solutions, 23
Micelle Configurations, 24
Micelles, 23
Micelles, Rod-Like, 23
Micelles, Spherical, 23
Microemulsion, 23, 40, 43
Microscopic Examination, 42, 44
Monodisperse Spheres, 30

Nature of Interfaces, 2
Nozzle Air Unit, 59

Oblate Form, 49
Occurrence, 64
Oil-in-Water (O/W) Emulsion, 4, 33, 41, 63
Oil Red XO Dye, 41
Oil-External Emulsion, 41
Oil-Soluble Dye, 41
Oleophillic, 21
One-Phase Region, 28
Origin and Occurrence, 105
Overall Planning, 81
Overtreat, 33

Packed Beds, 142
Parallel-Plate Separator, 74, 77, 142
Partial-Stream Pressurization Test, 57
Particle Charge Tester, 40
Particle Charge Testing, 52, 53
Particle Size Distributions, 44, 46, 47
Petroleum Emulsion, 105
Phase Diagrams, 24, 25, 27, 28, 29
Phase Rations, 31
Phase Reversal, 40, 83
Phase Rule, 25
Phipps-Bird Stirrer, 54
Plant Cells, 32
Polyhedral Shapes, 31
Polymodal Distributions, 51
Porous-Bed Coalescers, 141
Pressure Difference, 138
Production Zones, 105
Pseudoternary Diagrams, 27

Radius of Action, 10, 11
Random Walk, 18
Reverse Emulsion, 4, 125, 136, 138, 139
Reverse Osmosis, 138
Rhombic Dodecahedron, 31

Satellite Drops, 49
Scouring Effluent, 144
Sedigraph, 45
Sedimentation, 5
Settling, 5, 33
Settling Classes of, 33
  Type One, 33
  Type Two, 33
  Type Three, 34
  Type Four, 34
Settling, Hindered, 34
Short-Range Forces, 23
Skimmings, 93
Sludge Layer, 35
Sludges, 93
Sludge Treatment, 143

Smoluchowski-Muller Equation, 10
Sodium Decanoate, 4
Solubilization, 23
Specific Examples, 79
Spontaneous Emulsification, 51
Static Charge, 52
Steamfloods, 126
Stilling Ponds, 73
Stokes'Law, 11, 12, 15, 33, 45
Sudan Red, 111, 41
Suint, 144
Surfactant, 22

T-Cell Filter Separator, 141
Tertiary Recovery Methods, 126
Tessellate, 31
Testing, 39
Tests For Stability, 39
Tetrakaidecahedron, 31
Tetrahedral Regions, 32
Theory, 1
Thermodynamics, 25
Total Oil, 54
Total Oil and Grease, 55
Trash Trap, 67
Treatment Methods, 122
Truncated Octahedron, 31
Turbidity Measurements, 54
Turbullent Flow, 52
Two-Phase Region, 27
Types of Emulsions, 4
Types of Separators, 74
Typical Oil Field, 105
Typical Treatment Plant, 67

Ultrasonic Agitation, 51
Ultrafiltration, 102, 136, 139, 149
Unequal Spheres, 31

Variance, 25
Variability, 43
Viscosity, 31
Visual Examination, 39, 44, 54
Water-in-Crude-Oil Emulsions, 33
Water-External Emulsion, 40, 41
Water-in-Oil (W/O) Emulsion, 4, 41, 93
Water-in-Oil Emulsions Breaking Methods, 94
Water Knockouts, 109
Water Treatment Scheme Detailed, 69
Water Treatment Scheme General, 68
WEMCO Lab Flotation Machine, 57
Wet Oils, 63, 93
Wool Acid Cracking, 146
Wool Grease, 144
Wool Pulled, 144
Wool Raw, 144
Wool Sheared, 144
WRONZ System, 148

Yolk, 144